CONTENTS

Kirill Sergeev/Shutterstock

94 Making a Profit
The new entrants to supersonic passenger transport are financing their aircraft projects differently but will they make a profit?

102 Sustainable Supersonic
All sectors of aviation are under pressure to be more sustainable and supersonic aircraft are no exception.

110 Supersonic in 2050
Concorde was all about glamour and exclusivity, but to be successful supersonic passenger transport will need to be more inclusive.

Lockheed Martin

Original cover art by Antonis Karidis for Key Publishing
ISBN: 978 1 83632 149 1
Editor: Michael Doran
Senior editor, specials: Roger Mortimer
Email: roger.mortimer@keypublishing.com
Cover Design: Steve Donovan
Design: SJmagic DESIGN SERVICES, India
Advertising Sales Manager: Sam Clark
Email: sam.clark@keypublishing.com
Tel: 01780 755131
Advertising Production: Becky Antoniades
Email: Rebecca.antoniades@keypublishing.com

SUBSCRIPTION/MAIL ORDER
Key Publishing Ltd, PO Box 300,
Stamford, Lincs, PE9 1NA
Tel: 01780 480404
Subscriptions email: subs@keypublishing.com

Mail Order email: orders@keypublishing.com
Website: www.keypublishing.com/shop

PUBLISHING
Group CEO: Adrian Cox
Publisher: Steve O'Hara
Published by:
Key Publishing Ltd, PO Box 100,
Stamford, Lincs, PE9 1XQ
Tel: 01780 755131
Website: www.keypublishing.com

PRINTING
Precision Colour Printing Ltd, Haldane, Halesfield 1, Telford, Shropshire. TF7 4QQ

DISTRIBUTION
Seymour Distribution Ltd, 2 Poultry Avenue, London, EC1A 9PU
Enquiries Line: 02074 294000.

We are unable to guarantee the bona fides of any of our advertisers. Readers are strongly recommended to take their own precautions before parting with any information or item of value, including, but not limited to money, manuscripts, photographs, or personal information in response to any advertisements within this publication.

© Key Publishing Ltd 2025
All rights reserved. No part of this magazine may be reproduced or transmitted in any form by any means, electronic or mechanical, including photocopying, recording or by any information storage and retrieval system, without prior permission in writing from the copyright owner. Multiple copying of the contents of the magazine without prior written approval is not permitted.

www.key.aero 5

THE FIRST 50 YEARS

The First 50 Years

Supersonic passenger transport has had a chequered career after Concorde and the Tupolev Tu-44 launched, but the next decade may see its return.

WELCOME

Supersonic
Past and Future

I am writing this page from Colombia and I can only imagine what a difference supersonic flight would make to my laborious 50 hours of travel time home to Australia. Aboard Concorde, say, it would have taken little more than 12 hours.

The image of Concorde and supersonic travel was one of glamour, celebrities and ticket prices beyond the reach of most of the population, whereas the real value of supersonic aircraft to society must be about connecting nations and populations sustainably and in the most efficient way possible.

Almost everyone who saw Concorde in the air or at an airport was mesmerised by its graceful beauty and the awesome power and sound of its four Olympus 593 engines. This book is an early celebration of the 50th anniversary of Concorde's entry into service on January 21, 1976, and also a look forward to the coming decades when supersonic passenger airliners will once again appear in the skies.

Aspiring individuals and organisations can see opportunities to take advantage of new aeronautical technologies, advanced materials and a shift in political will to announce projects. Some projects have fallen by the wayside, but others are forming collaborations to share the knowledge and experience needed to turn their supersonic dream into reality.

In one case, US company Boom Supersonic is so well advanced that it will have a certified supersonic airliner ready to go by the end of this decade, while national research agencies work to resolve the sonic boom noise issue that has been in the 'too hard' basket since Concorde was launched.

The International Civil Aviation Organisation is working with agencies, including NASA, to gather the scientific data needed to set effective regulations on how much sonic boom noise reaches the ground and the Federal Aviation Administration has been ordered by the US President to rescind laws that unilaterally ban supersonic flights over land.

By turning the pages you will gain insights into this exciting world and the current aircraft and engine developments that will revive supersonic flight and finally allow new aircraft to operate over land. In essence, the future is about learning from Concorde and the Soviet Tupolev Tu-144, but improving what they achieved in an age where sustainability, cost and efficiency need to be in harmony.

As we near Concorde's 50th birthday, I hope you enjoy reading this as much as I did researching and writing it.

Michael Doran
Editor

▼ It's time for new dreamers to honour Concorde's supersonic legacy. NARA

CONTENTS

Travelview/Shutterstock

Contents

6 The First 50 Years
The 50th anniversary of the first commercial flights of Concorde is in January 2026 and we look back at the origins of supersonic passenger transport.

12 Concorde's Legacy
Concorde is far too quickly dismissed as a commercial failure, but many of the lessons learned formed the foundation of next-generation supersonic aircraft.

20 JAXA Boom Busters
Japan's Aerospace Exploration Agency is developing low sonic boom concepts in collaborations with multiple partners, including NASA and Boeing.

28 NASA X-59 Experimental
NASA has partnered with Lockheed Martin Skunk Works to develop the X-59 experimental aircraft that is testing concepts to reduce sonic booms.

40 Boom Time
In just over a decade, Boom Aerospace has built and flown a supersonic demonstrator and is on track to launch its Overture airliner by the end of the 2020s.

50 China Connection
China's leadership sees an opportunity to leapfrog Western aerospace giants with supersonic airliners.

58 Hits and Misses
Developing new aircraft or engines is not a game for the faint-hearted. We look at some successes and misfires in recent times.

68 Super Power
How the Olympus 593 engine that powered Concorde has influenced Boom Supersonic's bespoke Symphony and the GE Affinity powerplants.

78 Regulating the Boom
The ban on supersonic flight over land limited Concorde's success, but new standards and technology may surmount those issues.

88 Hypersonic Promise
It will be decades before hypersonic airliners become a reality, but there is significant work going on to build technology to make it happen.

Boom Supersonic

THE FIRST 50 YEARS

▲ Air France Concorde on the edge of Paris CDG in 2010. Senohrabek/Shutterstock

▼ Tupolev's Tu-144 operated just 55 scheduled passenger services. Kirill Sergeev/Shutterstock

Humans have long been captivated by speed. The land and water speed records of Sir Donald Campbell. Usain Bolt in the Olympic 100 metres. Or an F-35 roaring over an airshow crowd. The need for speed has been more visible in the development of aviation than almost any other human pursuit, allowing travellers to visit faraway places in ever shorter times.

Longer range jets, such as the Boeing 707, DC-10 and 747, brought any point on the planet within easy reach and transformed the way business and leisure travellers connected with family, friends and associates. With flights becoming longer and more tiring there was always the dream of connecting major cities such as London, New York, Sydney, Delhi, Paris, Singapore, Rio de Janeiro and Cape Town in just a few hours.

Perhaps the most challenging and ambitious aviation frontier has been the evolution of supersonic passenger airliners that could cross continents and oceans in a fraction of the time needed by subsonic aircraft, slashing long- and medium-haul flights by more than half.

Origins of supersonic flight

The evolution of supersonic passenger flight is a fascinating story of political ambitions, technological breakthroughs, project failures, economic constraints, environmental concerns and human tragedy. It had a stuttering post-war start, blossomed in the 1970s, fizzled out at the turn of the century and is now making a comeback, although the industry has yet to prove it can return in the sustainable or quieter way that will allow supersonic flights over land.

The pursuit of supersonic aircraft began in earnest in the 1950s, spurred on by the advances in high-speed flight achieved during World War Two and the early jet age, as well the rapid evolution of enhanced aerodynamics, propulsion systems and materials. What really captured the attention of engineers and aviation pioneers was when test pilot Charles 'Chuck' Yeager first broke the sound barrier aboard the Bell X-1 in 1947, proving that supersonic speeds were achievable.

On October 14, 1947, and at the age of 24, Yeager became the first pilot in history confirmed to have exceeded the speed of sound in level flight, while piloting the rocket powered X-1 *Glamorous Glennis* at Mach 1.05 (806mph) over the Rogers Dry Lake in the Mojave Desert in California. The aircraft he flew that day was later put on permanent display at the Smithsonian Institution's National Air and Space Museum in Washington, DC.

With breaking the sound barrier now a matter of record, aerospace engineers saw the opportunity to create aircraft that could carry passengers at twice the speed of sound and slash intercontinental travel times. Multiple projects were conceived, including the Boeing

THE FIRST 50 YEARS

▲ **A British Airways Concorde on display in the UK.** Just Jus/Shutterstock

2707 in the US. In the end, it was the rivalry between the Soviet Union and a United Kingdom/France partnership that kickstarted supersonic travel.

By the late 1950s and early 1960s, post-war national pride and Cold War competition spurred both Soviet and Western aerospace industries to pursue the supersonic quest as symbols of national progress. Early supersonic transport concepts emerged after military research into supersonic fighters and bombers provided much of the aerodynamic and propulsion knowledge needed for civil applications.

Supersonic Icons

While the Soviet Union and the Western alliance had already had their space race, they were soon competing to be the first to design, build and operate a supersonic transport aircraft. While it ultimately turned out to be a pyrrhic victory, it was the Soviets who made it to the finishing line first, with their Tupolev Tu-144 making its maiden flight on December 31, 1968, two months ahead of Concorde.

The Tupolev Tu-144 was designed to cruise at speeds of Mach 2 at 60,000ft and was intended to serve Aeroflot's high-speed long-haul routes and showcase Soviet technological progress. It was powered by four Kuznetsov NK-144 afterburning turbojet engines and featured a sleek delta wing design and retractable canards for improved low-speed handling.

It entered service carrying freight and mail and did not start passenger transport until 1977, but the aircraft was plagued by technical issues, high fuel consumption, excessive noise and safety concerns. The last of these was exacerbated by a fatal crash in 1973, which, coupled with the reliability issues, led to the Tu-144 retiring from passenger service in 1978 after just 55 scheduled flights.

On June 3, 1973, a Tupolev Tu-144 was performing a demonstration flight at the Paris Air Show, when it disintegrated in the air while performing extreme manoeuvres and fell on the town of Goussainville, Val-d'Oise, France. The crash destroyed 15 houses and killed all six people onboard the aircraft and eight more on the ground, including three children, as well as 60 people receiving severe injuries.

▶ **Chuck Yeager broke the sound barrier in this Bell X-1.** crbellette/Shutterstock

THE FIRST 50 YEARS

The incident occurred during the second demonstration flight of the day, when the Tu-144 performed a high-speed pass along the runway and then, with all four engines at full power, went into a steep and rapid climb. The aircraft stalled below 2,000ft and fell into a steep dive. As the pilot tried to pull out of the dive, the left wing outboard of the left-hand engines broke away and overload stresses fractured the fuselage forward of the wing.

Conversely, the Concorde, developed jointly by British Aircraft Corporation and Aérospatiale, was proclaimed an engineering marvel with its slender delta wings, drooping nose and afterburning engines that allowed it to cruise comfortably at Mach 2 (1,535mph) or more than twice the speed of sound at 60,000ft.

The Concorde project was announced in 1963 and entered service 13 years later on January 21, 1976, when British Airways and Air France launched simultaneous flights at 1140hrs. Initially, both airlines operated alternative routes, with Air France flying Paris-Rio de Janeiro via Dakar while British Airways flew London-Bahrain in just four hours, compared to more than six hours on subsonic aircraft.

Concorde had an illustrious career that spanned nearly three decades and, despite all the doom and gloom around its cost and operating economics, it actually produced profits for British Airways, although the different financial model used by Air France and the French government produced poor results across the Channel.

There was initial widespread interest in Concorde from major airlines, including American Airlines, Air Canada, Braniff, Pan Am, TWA, Qantas, Air India, Lufthansa and Iran Air, who placed purchase options that were later cancelled. Ultimately only 14 production aircraft entered service and the programme never fulfilled its early commercial promise, due in part to the ban on over land supersonic flights. The only other airline to participate in Concorde flights was Singapore Airlines, with a dual-liveried example co-operated with British Airways flying the Singapore-London route in the late 1970s and early 1980s.

While these two iconic aircraft had vastly different careers, both airplanes played their part in awakening the aerospace community and the travelling public to how supersonic aircraft could open new possibilities in flight.

Supersonic hibernation

After the fanfare and excitement of Concorde and the Tupolev Tu-144 entering service, the era of supersonic flight faltered and for decades no new supersonic passenger aircraft reached production.

In the US, the Boeing 2707 programme, which aimed to deliver a Mach 2.7 aircraft, was cancelled in 1971 due to spiralling costs, environmental concerns and an overall lack of interest from major airlines. In discussions around supersonic flight the detrimental

◀ The Tupolev Tu-144 had a short and eventful career with Aeroflot. Yuliya D'yakova/Shutterstock

▼ Boom's Overture is likely to be the next supersonic passenger airliner. Boom Supersonic

THE FIRST 50 YEARS

▲ JAXA is developing a 10m unmanned vehicle for testing. JAXA

impact of sonic booms and regulations banning flights over land comes up time and again.

A scan of the airlines that placed purchase options for Concorde shows that many of them – such as Air India, Iran Air, Lufthansa, Qantas and the major US carriers – needed an aircraft that could fly at supersonic speeds over land to make the economics and high operating costs financially viable.

The operational realities of supersonic flight proved sobering, particularly as the fuel consumption of both Concorde and the Tu-144 was enormous, especially during supersonic cruise. Adding to this, the oil crises of the 1970s drove up costs, noise around airports came under increasing scrutiny and a growing environmental movement began to question the wisdom of such energy-intensive travel.

The devastating crash of an Air France Concorde in 2000 shocked the world, and although British Airways made design changes to improve safety and carried on flying for a few more years that tragedy, along with spiralling maintenance costs, virtually pulled the curtain down on supersonic passenger transport.

Back to the future

The 2020s have seen a revival of interest in supersonic passenger transport. In 2025, several companies and aerospace institutions are actively engaged in the quest to diminish the impact of sonic booms and enable supersonic airliners to legally fly over land.

Today's focus is on overcoming the limitations that so negatively impacted earlier designs and aircraft, including Concorde, the Tu-144 and Boeing's 2707 program. Engineers are developing quieter propulsion technologies to mitigate the sonic boom, using computational modelling and experimental designs to reshape shock waves.

Fuel efficiency and environmental impact are at the forefront of development, with an emphasis on sustainable fuels and reduced emissions that will apply to all commercial aircraft, including supersonic airliners. Concorde was allowed to operate with special dispensations in that regard, but the International Civil Aviation Organization (ICAO) has made it clear that the new standards it is working to develop will be universal.

A variety of projects have been talked about as supersonic concepts, but by late 2025 only one candidate seems likely to get a supersonic airliner in the air and ready to commence operations in the next decade. From its headquarters in Colorado, Boom Supersonic has evolved from a vision of its founder Blake Scholl in 2014 to having a supersonic aircraft demonstrator that is gathering the data and experience that will underpin the company's Overture supersonic passenger jet.

Overture is a 65-80 seat supersonic airliner that is targeting Mach 1.7, which is around twice the speed of today's subsonic airliners, on more than 600 routes. Boom is developing a Boomless Cruise concept that uses Mach cut-off principles where the sonic boom refracts in the atmosphere and never reaches the ground. The effect is achieved by breaking the sound barrier at a high enough altitude, with exact speeds varying depending on atmospheric conditions, which Boom will achieve by using real-time weather data on advanced computers. The bespoke Symphony engine enhances transonic performance compared to commercially derived powerplants, allowing Overture to efficiently transition to supersonic speeds at altitudes above 30,000ft.

Overture will also be fitted with an advanced autopilot that uses real-time weather data and software algorithms to automatically select the optimal speed for Boomless Cruise. While that may sound a little theoretical, Boom

▼ Boom's XB-1 demonstrator has broken the sound barrier with no sonic boom reaching the ground. Boom Supersonic

THE FIRST 50 YEARS

◀ Singapore Airlines operated a dual-liveried Concorde with British Airways. airteamimages.com

has already achieved Boomless Cruise on six supersonic flights of its XB-1 demonstrator over the Mojave Desert.

There are two other projects that are purely experimental aircraft designed to provide regulators and manufacturers with data that ICAO will use to set new sonic boom over land regulations. These involve NASA's X-59 experimental aircraft and a supersonic concept under development by the Japan Aerospace Exploration Agency (JAXA,) but neither of these involve a new commercial supersonic transport aircraft.

NASA's X-59 Quiet SuperSonic Technology (Quesst) project is a key initiative that is aimed at setting new noise standards by reducing sonic booms typically caused by supersonic flight. A primary goal is to enable commercial supersonic flights over land through the use of the X-59 experimental aircraft, a piloted single-engine jet built by Lockheed Martin Skunk Works.

Formed in 2003 JAXA is Japan's national research and development agency specialising in space and aeronautics. In 2024, it launched the Re-BooT program to design a concept of a quiet supersonic transport aircraft and to demonstrate its robust low-boom design technology through flight tests.

The primary aim of JAXA is to design and validate robust low-boom aircraft shapes that consistently minimise sonic boom intensity and ensure that noise can be reduced across a wide range of flightpaths, altitudes and atmospheric conditions. For this programme, JAXA will produce an unmanned experimental vehicle approximately ten metres long that will be mounted on a manned aircraft. At an altitude of around 42,000ft, it will separate and use gravity to accelerate to supersonic speed, with onboard computers controlling the vehicle and directing it to fly over a measurement system on the ground. The agency is using Re-BooT to bridge the gap between fundamental research and real-world application, while also bringing supersonic travel with noise levels acceptable for over-land operations closer.

Concorde and the Tu-144 are now museum pieces, but their legacy lives on in this renewed quest for faster, greener and more accessible air travel. The next five to ten years will reveal if supersonic passenger transport is returning or if the dream is again dashed and put in the 'too hard' basket.

▼ Qantas was one of many airlines to place options for Concorde. Qantas

www.key.aero 11

CONCORDE'S LEGACY

Concorde's
Legacy

Concorde was ahead of its time and the technologies and excitement it created live on as the platform for the revival of supersonic passenger transport.

A legacy is built when a product endures well beyond its initial launch and transcends time, trends and fleeting consumer interest. The influence of a legacy product can be measured by how it shapes industries, drives technological advancement or becomes a cultural icon.

Leaving a legacy is about contributing something meaningful that continues to add value, spark creativity and foster progress long after its creators have moved on and is a mark of a vision fulfilled and purpose achieved. Legacies are built by solving real problems and connecting emotionally with users to become woven into the fabric of everyday life.

The designers, builders and operators of Concorde created all of that with an aircraft that has endured for nearly 50 years after it first entered commercial service on January 21, 1976, and well beyond its final commercial flight was operated by British Airways on October 23, 2003.

What is unusual is that it has taken more than 20 years for any real interest to emerge that follows on from Concorde's legacy. Assuming Boom Supersonic becomes the first company to resume commercial supersonic services, it will be closer to 30 years for the quest to begin again.

To highlight Concorde's legacy, the long list of innovations has been grouped under the following five headings: technology, passenger experience, engines and collaborations, future prospects and emotional impact.

Technology

This is the area where Concorde really stands out, as its designers and engineers continually solved significant challenges on the run and devised solutions that were well ahead of anything that was available at the time.

The slender delta wing, ogee planform and droop nose added so much to the grace of the aircraft, but were also carefully refined solutions to the aerodynamic challenges of sustained supersonic flight. Concorde also pioneered fly-by-wire (FBW) control augmentation in civil aviation, and although it was not a full digital FBW, it had mechanical controls assisted by electronic and hydraulic systems to maintain stability in thin air at high altitudes.

The experience in control law development for a slender delta aircraft at high speed informed later electronic control systems for high-performance fighters, with Dassault's Rafale a beneficiary. The Rafale's delta-canard configuration benefits from this lineage, with relaxed stability made manageable through digital flight control, a concept to which Concorde's engineers contributed key aerodynamic data.

Engineers had to find advanced materials that could cope with the extreme heat at Mach 2 that would cause the airframe to expand by up to 25cm. This required advanced aluminium alloys and precision engineering to maintain Concorde's structural integrity.

Many of Concorde's design principles informed later developments in military and experimental aircraft and its precision manufacturing techniques influenced commercial aviation standards. One example is that Concorde's delta wing with ogee planform was designed to maintain lift at both subsonic and supersonic speeds without excessive drag. Pure supersonic fighter jets are optimised for short bursts of speed, whereas Concorde needed efficiency in the long-range cruise at Mach 2. The understanding of vortex lift generated by the wing's leading

◀ **Air France Concorde with gear extended.** Herget Josef/Shutterstock

CONCORDE'S LEGACY

▲ Concorde at Hermeskell Airport in Germany. Joseph Creamer/Shutterstock

edge was directly applicable to military aircraft such as the Dassault Rafale and the Eurofighter Typhoon, with both exploiting controlled vortices to enhance manoeuvrability.

A second example is that heat management at sustained supersonic flight posed unique thermal challenges. Concorde's skin heated to more than 120°C, which caused the fuselage to expand by up to 25cm, and the use of high-strength aluminium alloys and careful expansion joint engineering informed later supersonic and hypersonic projects. The SR-71 Blackbird, a former military reconnaissance platform, used titanium, but the structural expansion and contraction principles of Concorde were applicable to composite-metal hybrids in experimental high-speed jets. Concorde also provided data on fatigue cycles under thermal stress, influencing long-term structural integrity planning in aircraft operating at high Mach numbers.

Passenger experience

Despite relatively few people ever experiencing what it was like to travel on Concorde, the mere mention of its name evokes feelings of glamour, exclusivity, prestige and celebrity which will probably never be replicated by any of the new entrants to supersonic passenger transport.

For those who could afford a ticket, Concorde redefined what air travel could mean. Even with its narrow cabin carrying just 92 to 128 passengers, there was nothing else like flying on Concorde. We have to keep remembering that this was an aircraft designed in the 1960s and 1970s, so the distinctly business-class feel was a novelty long before such concepts gained popularity 20 years later.

When Concorde launched services, a flight between London and New York took around 7-8 hours, but the new supersonic service could do the same flight in just three-and-a half. Cruising at 60,000ft allowed passengers to see the curvature of the Earth and the deep blue of the stratosphere, and flying on Concorde often meant rubbing shoulders with celebrities, famous sportspeople, business tycoons and heads of state.

The hype and aura surrounding Concorde influenced the aspirations and marketing of other aircraft manufacturers and airlines, planting the idea that using new technology could transform air travel from being mundane and something to be endured to a magical experience.

Today airlines talk incessantly about the passenger experience. Usually that means the food, beverages, seats, amenities, entertainment and so on at the pointy end. Concorde itself was the passenger experience and stood alone because of that. Although it had its share of premium comfort, it really didn't need it to provide the greatest passenger experience commercial aviation has ever seen.

Engines and collaborations

Another technical marvel was the Rolls-Royce/Snecma Olympus 593 turbojet that relied on reheat capability (afterburners) to deliver

CONCORDE'S LEGACY

▲ There's plenty to look at in a Concorde cockpit. Antonis Kousoulas/Shutterstock

the significant thrust needed for supersonic cruise and take-off from relatively short runways.

In terms of an enduring legacy, the sophisticated intake control systems needed to optimise engine airflow at high speeds make the engine intake design one of Concorde's most valuable contributions to aerospace engineering. The Olympus 593 turbojets required a precise airflow regime and, at Mach 2, intake air had to be slowed from supersonic to subsonic speeds before entering the compressor. Concorde's computer-controlled, variable-geometry intakes used moving ramps and spill doors to manage shockwaves and prevent compressor stalls.

This principle was later refined in high-speed military jets, including the Eurofighter Typhoon and Panavia Tornado, which adopted advanced variable geometry inlets for optimal engine performance across a wide speed range. Concorde also pioneered the

◀ The Rolls-Royce/Snecma Olympus 593 turbojet. Tom Meaker/Shutterstock

CONCORDE'S LEGACY

▶ Concorde on display in New York City. Phil Emmerson/Shutterstock

▼ Concorde G-BOAB on take-off climb. John Selway/Shutterstock

discipline of intake shockwave management, which has featured on experimental aircraft such as NASA's X-43 and X-51 hypersonic test vehicles.

Another aspect of supersonic technology is that Boom Aerospace has declined turning to the major engine OEMS for a powerplant for its Overture supersonic airliner. Instead Boom opted to form a collaboration with a small group of expert partners to build its own bespoke engine, Symphony, and with the engine already under construction the results have been outstanding. The relevance to Concorde is that this model of co-operation and collaboration was not common in the 1960s and the supersonic programme was one of the earliest large scale joint aerospace projects between two major European nations. The Anglo-French partnership was probably one of the tightest contracts in existence, as neither could withdraw without the other's consent and both shared the risks and the costs.

The Concorde partnership proved effective and set the model for future co-operative ventures, most notably the formation and development of Airbus. Many engineers, designers and managers who brought Concorde to life later contributed to Airbus projects such as the A300 and A320, bringing with them valuable experience in joint development.

Future impacts

For aviation, the last five years of this decade are shaping up to be dominated by the revival of supersonic passenger flights, with the challenges that limited Concorde's commercial success now being addressed nearly 50 years after the iconic aircraft entered service.

Some of the world's finest aerospace minds at organisations such as NASA and Japan's Aerospace Exploration Agency (JAXA) are working on reducing the impact of sonic booms reaching the ground and ICAO is developing scientifically based noise limits instead of the current blunt instrument of lawmakers simply banning supersonic flights over land.

The impact of this flurry of activity is that new entrants to supersonic passenger transport will have a set of regulations that posit clear noise limits over communities and around airports that must be met. They will also have the benefit of the extensive research gathered by NASA's X-59 experimental aircraft that is the backbone of the quiet supersonic programme Quesst.

Concorde's difficulties surmounting these issues have educated regulators, prospective OEMs and airlines that the environmental and societal challenges need to be part of project, not something to be left to outside agencies to sort out when the aircraft is ready to fly.

Another aspect is that in today's political climate there is a strong impetus to lead global aerospace development, and politicians in the US are introducing new laws to help companies such as Boom Aerospace get their supersonic

▲ There are 16 Concordes on public display.
Colinmthompson/Shutterstock

CONCORDE'S LEGACY

▲ Even the tailfin on Concorde was elegant. Graham Taylor/Shutterstock

aircraft off the ground. US President Donald Trump has also joined in by ordering the US Federal Aviation Administration to end the ban on supersonic flights over land, subject to setting new noise limits. It is fair to say that without Concorde's legacy this level of intervention may not have been forthcoming.

A more direct link is that every supersonic commercial programme since Concorde's retirement has drawn comparisons to it, and the respect for the original sets the benchmark by which any new aircraft will be judged. Concorde's long shadow has set Mach 2 cruise at 60,000ft as the baseline for new supersonic airliners, while its decades of real-world supersonic operations will provide airlines with a blueprint of how to run their services.

Emotional legacy

Following in the footsteps of such an icon as Concorde is a two-edged sword for all potential new operators, including Boom Supersonic and Spike Aerospace. The sheer excitement that surrounded Concorde will create a storm of public interest and media attention for new entrants, but they will have to live up to the high expectations that people have formed based on what Concorde delivered.

▶ Avionics and instrumentation in the Concorde cockpit. Jurav/Shutterstock

CONCORDE'S LEGACY

▲ Air France Concorde at Paris/Charle de Gaulle Airport. travelview/Shutterstock

As has happened with the much-loved Boeing 747, the physical presence of Concorde has lived on despite the aircraft being out of service for more than two decades. It may surprise some to know there are 16 surviving Concordes on display in museums and at airports across the United Kingdom, France, Germany and the United States (with another in storage in Barbados) that attract thousands of visitors annually.

While most who tour the aircraft will be aviation enthusiasts, there will many more who are there to tour what is probably the world's most recognisable and futuristic passenger aircraft. At a time when aviation is scrambling to attract new pilots, technicians and ground staff, a school trip to visit Concorde will kindle the enthusiasm in some young visitors to see aviation as a career choice.

The emotional legacy is perhaps the strongest pull of all, as anyone who flew on or witnessed firsthand the aircraft take off will always remember the sheer noise and ground-trembling power of the Olympus 593 turbojets as it thundered into the sky. Or at the end of a flight, the way the aircraft glided in to land, resembling some mythical bird slowly surveying its prey from above and gliding down to Earth.

There has never been another aircraft that can evoke such excitement and public attention. When the Boom Supersonic Overture enters service in the early 2030s, people will clamour to get a sight of it thanks largely to the memory and legacy of Concorde.

While today's aircraft are extremely efficient and reliable, there is a certain sameness around airports these days. And although the Airbus A380 is a joy to watch, no other aircraft can stop people in their tracks as soon as it appears on the runway or approach.

The intangible sense of wonder is missing from commercial aviation and the arrival of Overture will be the next time that feeling will return.

More than 20 years after its retirement, Concorde's legacy lives on in so many ways and its influence is seen in every modern supersonic concept, in the collaborative workings of Airbus and the travelling public's imaginations. It has spawned many new technologies and aircraft advancements, while setting the benchmark for the current renewed interest in supersonic passenger transport. Concorde may not have been to everyone's liking and, as a product of the 1960s, its environmental footprint may not meet today's standards, but it was an outstanding success and is a testament to the aviation spirit shown by pioneers such as the Wright Brothers all those years ago.

 DON'T MISS OUT ON OTHER KEY AVIATION MAGAZINE SPECIALS
If you'd like to be kept informed about Key Publishing's aviation books, magazine specials, subscription offers and the latest product releases. **Scan here »**

JAXA Boom Busters

The Japan Aerospace Exploration Agency's groundbreaking work on sonic boom mitigation is providing scientific rigour to set new supersonic noise limits.

▶ The sleek lines of the low-boom aircraft used in D-SEND #2. JAXA

The Japan Aerospace Exploration Agency (JAXA) is a leading research and development agency specialising in space and aeronautics that conducts a wide range of research at more than ten centres across the country. It was formed in 2003 and is a key part of Japan's space exploration, satellite development and aerospace innovation efforts.

JAXA's mission is to help create a safe and prosperous society through its focus on key fields such as space exploration, human spaceflight, satellite development and aeronautics research. It is a major partner in the International Space Station and is developing the Smart Lander for Investigating the Moon programme.

JAXA leads research into quiet supersonic aircraft, electric propulsion, hypersonics and next-generation air traffic management, including research and development projects that contribute to the progress of supersonic aircraft both in Japan and worldwide. It recognises that momentum for developing supersonic transport has been increasing in the last decade and that the International Civil Aviation Organization (ICAO) is looking to set new standards for supersonic flights over land.

Silent supersonic

Between 2016 and 2020, JAXA's S4 project focused on the research and development of a conceptual aircraft for supersonic passenger transport based on assumptions of Mach 1.6, 50 passengers, an approximately 70 tonne take-off weight and a cruising distance of 3,500nm or more. The key technical goals included:

- **Sonic booms** To enable supersonic flight even over land, the sonic boom intensity must be 85PLdB or less.
- **Take-off and landing noise** Noise at airports should meet ICAO Chapter 14 standards as applied to current subsonic passenger aircraft.
- **Air resistance** To reduce fuel consumption and increase cruising range, air resistance needs to be reduced to achieve a cruise lift-to-drag ratio of 8.0 or higher.
- **Weight** To reduce fuel consumption and increase cruising range, the structural weight must be at least 15% less than that of Concorde.

JAXA BOOM BUSTERS

JAXA has provided a very clear description of how sonic booms are created, explaining that when aircraft operate at supersonic speeds shockwaves are generated from various parts of the jet that integrate as they propagate over long distances in the atmosphere. That creates an N-shaped pressure waveform that causes two abrupt pressure fluctuations on the ground or a sonic boom that is characterised by a loud, impulsive sound.

The challenge to mitigate or prevent sonic booms reaching the ground is a recurring theme throughout this bookazine and every aspiring supersonic start-up needs to find a solution or be confined to flying at subsonic speeds over land.

The ICAO is working to develop international standards for sonic boom regulations and JAXA is contributing valuable research by providing the results of its D-SEND project and other evaluations testing human perception of sonic booms using a simulator, a small booth that reproduces the same sound pressure changes as the actual sonic boom using low-frequency loudspeakers mounted on the walls. The JAXA simulator uses eight large low-frequency and four small high-frequency loudspeakers to play back the simulated sonic boom to subjects to evaluate sonic boom sounds with various waveforms.

D-SEND project

D-SEND is a multi-phase experimental programme within JAXA's supersonic research and development efforts, with a core mission to demonstrate low sonic boom design concepts via flight testing and to develop aerial measurement technology to record sonic boom waveforms for data-driven design and regulation. The full project name is Drop Test for Simplified Evaluation of Non-symmetrically Distributed Sonic Booms.

It has two phases and, in D-SEND #1, two different axisymmetric bodies are dropped and the sonic booms from them are measured and compared. D-SEND #2 uses an unmanned, engineless experimental supersonic airplane that is based on JAXA's low sonic boom design

▼ JAXA D-SEND #2 flight model. JAXA

JAXA BOOM BUSTERS

▲ Wind tunnel testing for the JAXA D-SEND #2 aircraft.
JAXA

▶ Sonic booms were measured using tethered balloons.
JAXA

▼ JAXA is at the forefront of sonic boom mitigation.
JAXA

technology. In each phase, sonic booms are measured using an aerial measurement system with microphones installed along the line of a tethered blimp. The tests are performed at the Esrange Space Centre in Sweden, which provides a safe testing environment and balloon controlling techniques necessary to carry test bodies.

Between 2010–2011, JAXA performed experiments to confirm the functions of the prototypes used for measurement and recording units of the Airborne Blimp Boom Acquisition (ABBA) system and to acquire an actual sonic boom waveform for research on evaluation technology.

A boom measurement system (BMS) uses a low-frequency microphone that is moored at an altitude of 750m using a small balloon to prevent the influences from atmospheric turbulence above the ground on the sonic boom waveform. Microphones are placed on a mooring rope and on the ground to take measurements of changes in waveforms in the altitude direction, while meteorological data such as temperatures above the ground and at the altitude of the small balloon are also acquired.

The D-SEND #2 drop was conducted in Sweden in 2015, when the low-boom aircraft was released from a balloon at 30km altitude and flew over the measurement system at Mach 1.3 with a flight path angle of -50°. The test successfully measured low sonic boom signatures, which validated the design technologies used in the aircraft's nose and tail.

Robust low boom technology

Since the D-SEND project demonstrated the concept of low sonic booms, JAXA acquired technology to reduce the strength of the sonic boom to half that of conventional supersonic passenger aircraft. In 2024, JAXA launched the Robust En Route Sonic-Boom Mitigation Technology (Re-BooT) demonstration project to demonstrate robust low boom design technology through flight tests and to design a concept aircraft of a quiet supersonic civil transport.

JAXA BOOM BUSTERS

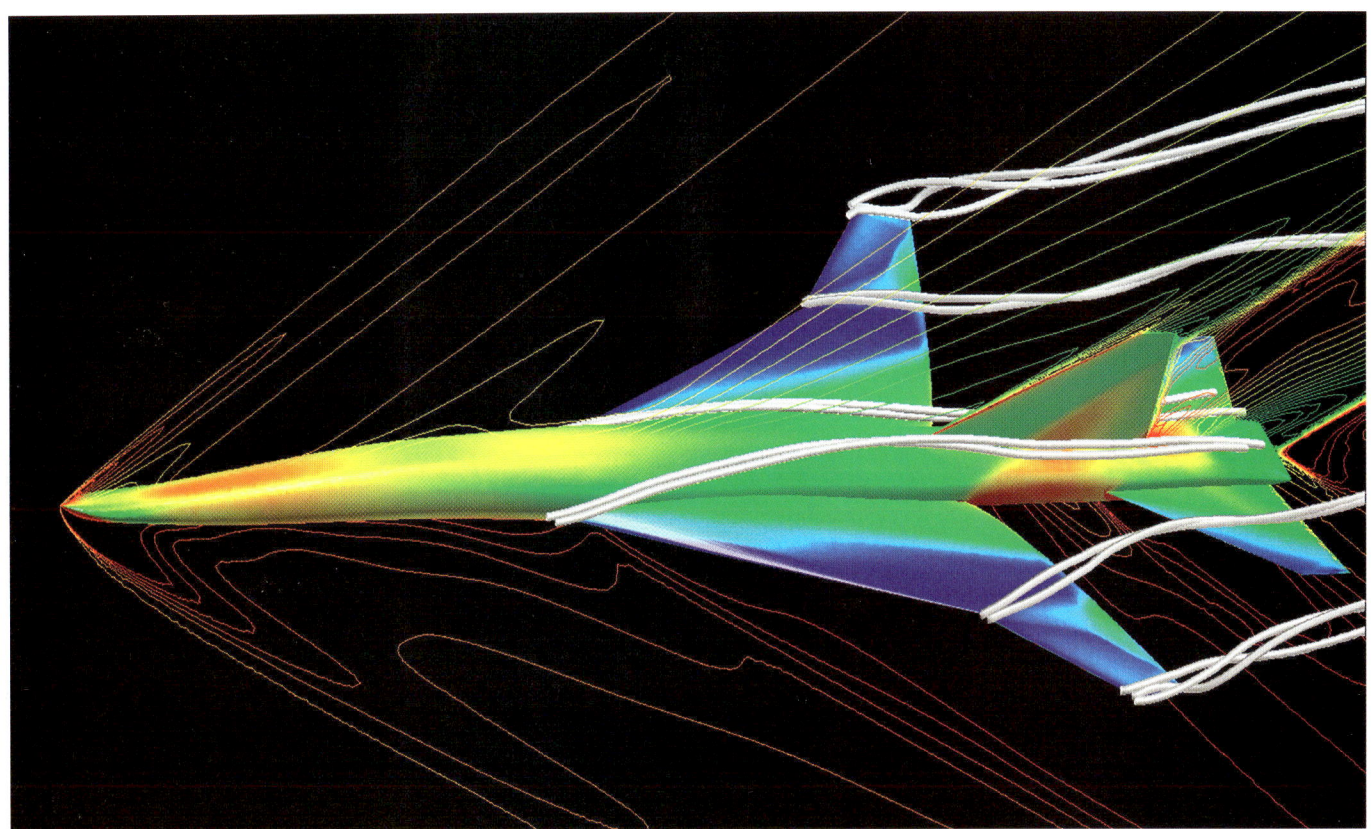

▲ Flight result analysis for the D-SEND #2 aircraft.
JAXA

Sonic booms do not only occur at the moment an aircraft exceeds the speed of sound, but all the time it is flying at supersonic speeds. The boom spreads conically from the aircraft and the ground location where it is heard moves along the supersonic flight path in a strip, otherwise known as a boom carpet. JAXA's Re-BooT reduces sonic booms observed throughout the boom carpet. The low boom design technology demonstrated in D-SEND was designed to reduce the sonic boom directly under the flightpath (on-track position) and under the design cruise speed condition (on-design condition). The technology to be demonstrated in the Re-BooT project is to reduce sonic boom in conditions other than design cruise speed (off-design conditions) and on the side of the flightpath (off-track position) and has been patented in Japan and overseas.

The Re-BooT programme aims to demonstrate that it's possible to make all sonic booms heard on the ground quieter. The central objective is to design and validate robust low boom aircraft shapes that consistently minimise sonic boom intensity under actual flight conditions. The earlier tests showed that sonic booms could be reduced under ideal circumstances and the Re-Boot program is focused on ensuring performance across a wide range of flight paths, altitudes and atmospheric variations.

◀ Centre of gravity measurements are carefully checked.
JAXA

JAXA BOOM BUSTERS

Near the ground, sonic boom waveforms are deformed by the unstable atmosphere, so to ensure the low boom design is accurately validated, a microphone will be hoisted using tethered balloons to an altitude where the atmosphere is less turbulent. In October 2024 JAXA said it planned to conduct these flight tests around 2028.

JAXA is involved in many collaborations on aerospace projects and this flight demonstration will be carried out as part of the research and development project funded by the Japan Science and Technology Agency (JST) under the Key and Advanced Technology R&D through the Cross Community Collaboration Programme (K Programme) established under the leadership of the Japanese government.

A key collaborator in the K Programme is Mitsubishi Heavy Industries (MHI), which is heavily committed though its co-operation with JAXA to the demonstration flight that will see JAXA design the shape of the aircraft using low-boom design technology, while MHI will take charge of the development and manufacture of the demonstration vehicle. As well as reducing the sonic boom, the actual passenger jet must meet all the usual certification criteria, including safety standards, and have a cruising range that is suitable for commercial flights.

JAXA Aviation Symposium 2024

The JAXA Aviation Symposium 2024 was held in Tokyo on October 18, 2024, with a focus on supersonic technology research and international collaboration. Experts from NASA and Boeing gave presentations, while a JAXA project

▲ D-SEND testing was done at the Esrange Space Centre in Sweden.
JAXA

For this, JAXA will design and produce a new experimental unmanned vehicle approximately ten metres long. Although engines will not be installed, an airframe will be designed with engine containers to evaluate the sonic boom waveforms generated by an aircraft shaped similar to a modern passenger jet. The experimental demonstrator will be mounted on a manned aircraft and, after climbing to an altitude around 42,650ft, it will separate and accelerate to supersonic speed by gravity. An onboard computer will autonomously control the aircraft and direct it to fly over the measurement system installed on the ground under specified conditions such as position and speed.

▶ One of the balloons used to measure sonic booms in Sweden.
JAXA

JAXA BOOM BUSTERS

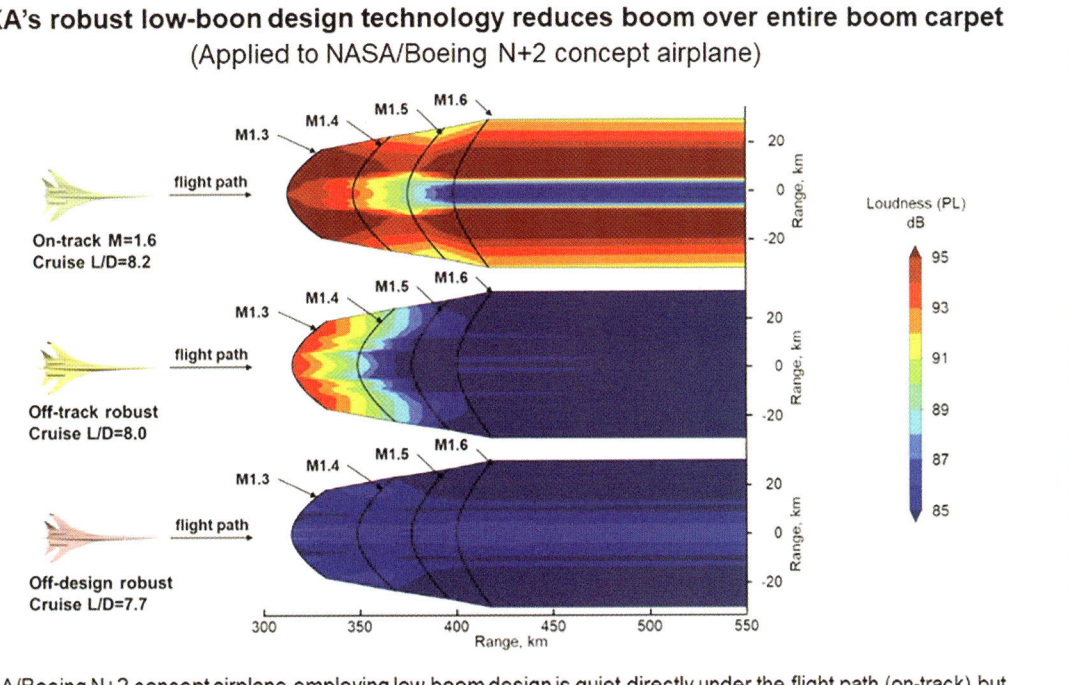

◀ JAXA'S low-boom technology works across the sonic boom carpet. JAXA

manager introduced the Re-BooT programme.

Carol Carrol, NASA deputy associate administrator for the Aeronautics Research Mission Directorate, made a presentation on the NASA X-59 research aircraft that is the centrepiece the agency's Quiet Super Sonic Technology (Quesst) mission. Carroll said the purpose of the programme was to gather data that would provide regulators with information that could help lift current bans on commercial supersonic flights over land. The long overdue first flight of the X-59 will follow decades of research with international partners, including JAXA and Boeing, that has found that each part of the aircraft, including the nose, canopy, tail and engine inlets, creates a shockwave travelling at different speeds that then merge.

Carroll told the audience that the very long, thin nose and unusual shape on the bottom of the X-59 does create shock waves, but they do not merge, so the sound heard on the ground is a much quieter 'thump' rather than a sonic boom, although NASA is aiming for the sound to be more like a car

▼ The Re-BooT programme flight demonstration test outline. JAXA

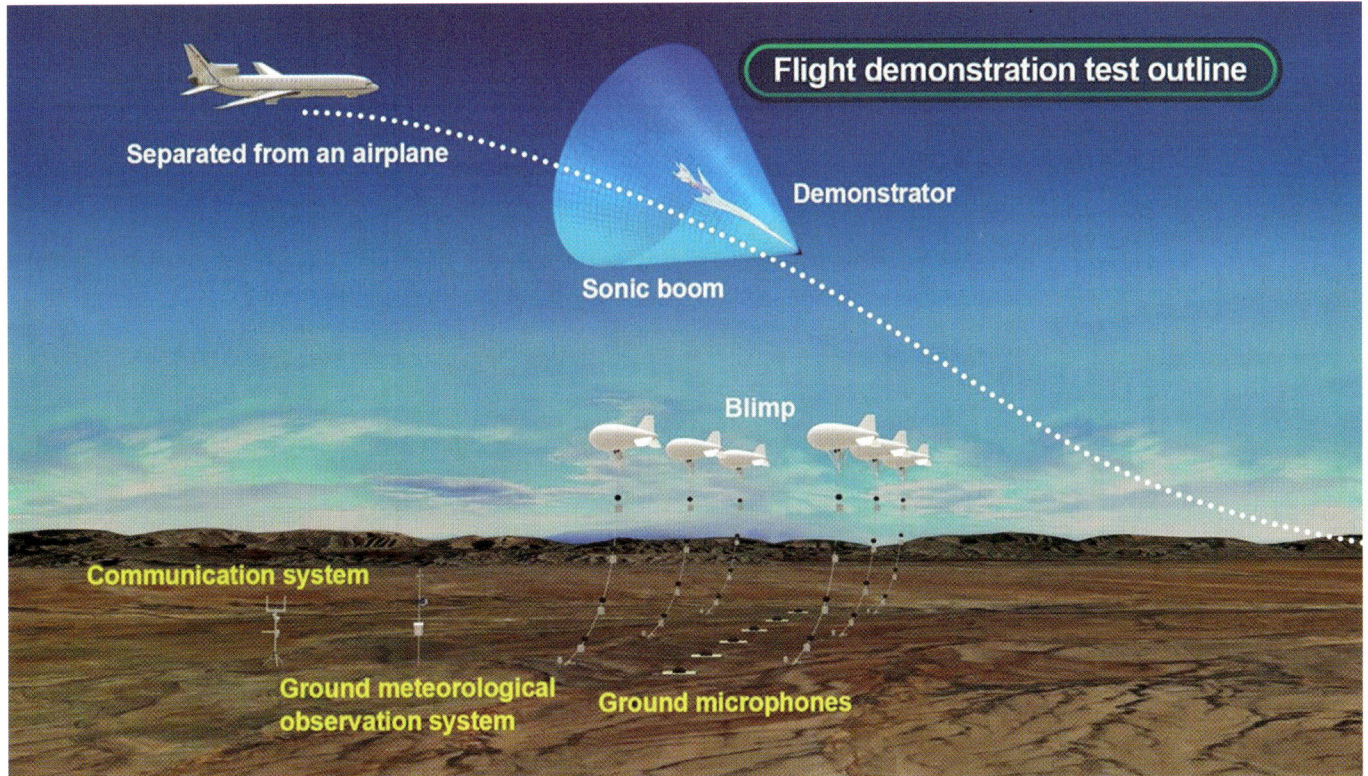

JAXA BOOM BUSTERS

▶ JAXA's low-boom technology effectively mitigates sonic booms. JAXA

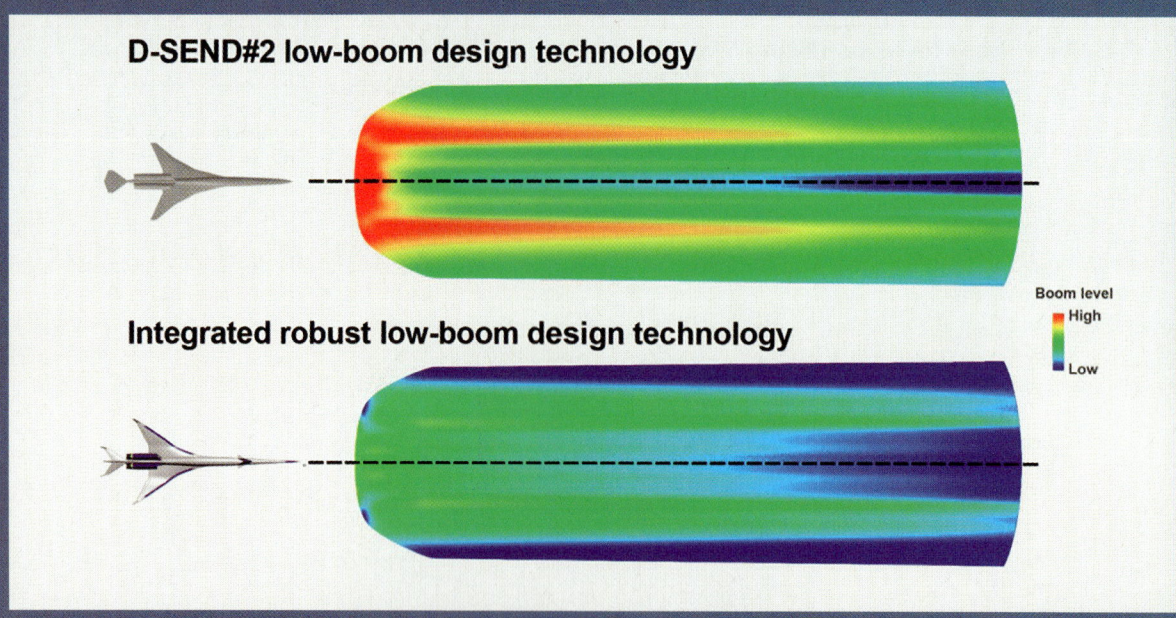

▶ One of the two different axisymmetric bodies used in D-SEND #1. JAXA

▼ Measuring the sonic boom with sophisticated listening devices. JAXA

door slamming in a neighbour's driveway.

The Quesst mission has three phases and NASA is still in the first phase of building the aircraft, before moving on to acoustic validation, which involves measuring acoustic pressure waves to verify that the shock waves are as designed. In phase three, the X-59 will fly over various communities in the US and overseas to collect data that will be provided to ICAO and the Committee on Aviation Environmental Protection (CAEP) to help regulators understand how to revise standards. To secure scientifically valid data, it will be collected from 10,000 to 100,000 points depending on the survey method employed, to determine if the sound level is acceptable

All prototypes are subject to rigorous testing before flying. JAXA

Wind tunnel testing for the XB-69 supersonic jet. JAXA

for people on the ground. NASA is planning to fly the X-59 up to six times a day to ensure enough test time to understand the effects of repeated exposure to the sound.

At the symposium, Dylan Jones, Boeing's executive director of Research & Technology Japan and the Boeing Korea Engineering & Technology Center, themed his presentation around global technology and innovation. He said that Boeing spends more than $3b annually on several initiatives to achieve its goal of decarbonising the aerospace industry by 2050.

Boeing is using four strategies to achieve its goal, including airline fleet renewal, improving operational efficiency, using renewable energy through sustainable aviation fuels, electrification and hydrogen and deploying advanced aircraft technologies. While most of those are based around sustainability, it is the last strategy that has special relevance to supersonic passenger transport. Boeing is engaged in various projects with JAXA, including the X-59, and the longest of those is the collaboration on supersonic flight, which began in 2006 and consists of four work packages: design, computational fluid dynamics (CFD) prediction workshops, wind tunnel testing and practical low-boom demonstration.

In April 2024, Boeing Research & Technology opened its Japan Research Center in Nagoya, which focuses on technologies for sustainability. Its work includes the use of digital tools for design, the development of lightweight carbon fibre, processes and applications for recycled carbon fibre and the investigation of the feasibility of hydrogen fuel-cell systems.

Most of what Yoshikazu Makino, manager for the ReBooT project, presented at the symposium has already been covered above. However, he did talk about Boeing's involvement in ReBooT, which was referred to as "collaborating on the design of a technical reference aircraft called the Low-Boom Supersonic Technology Concept Aeroplane." He said that, by applying JAXA's robust low-boom design technology to the NASA/Boeing N+2 concept aircraft, the resulting model will be provided to the ICAO for the standard development process.

Boeing's N+2 program is a conceptual research initiative aimed at developing next-generation commercial aircraft that would enter service after 2035. The 'N+2' name refers to aircraft two generations beyond the current fleet, with a focus on significantly improved environmental performance, noise reduction and fuel efficiency.

In partnership with NASA, Boeing Research & Technology division has evaluated several design concepts under the N+2 project, including blended wing body configurations and ultra-high bypass ratio engines. Supported by NASA's aeronautics research programmes, Boeing's N+2 studies have explored futuristic airframe designs, propulsion integration and advanced materials.

A key point that is sometimes overlooked is that JAXA is not designing a new supersonic airliner or entering the race to revive supersonic passenger transport. It is designing and developing concepts and experimental prototypes to study and validate its research into low-boom supersonic flights. This work is critical as aviation regulators and governments grapple with the dilemma of removing bans on supersonic flights over land, a move that is vital to the commercial success of commercial supersonic operations. JAXA is providing scientific rigour to assist ICAO to set new noise limits and its groundbreaking work on sonic boom mitigation is world class.

Preparing for the D-SEND sonic boom testing. JAXA

NASA X-59 EXPERIMENTAL

NASA X-59 Experimental

NASA is working with Lockheed Martin Skunk Works to develop the X-59 experimental airplane that will help set noise limits for supersonic flights over land.

NASA X-59 EXPERIMENTAL

Officially known as the X-59 Quiet SuperSonic Technology (Quesst) project, the NASA X-59 programme is a key initiative to dramatically change air travel by reviving supersonic passenger transport. It is also about setting new standards by reducing the sonic booms typically associated with supersonic flight.

Under the wing of the National Aeronautics and Space Administration (NASA), the programme has the goal of enabling commercial supersonic flights over land, something that has been banned in the United States since 1973 due to noise concerns. The centrepiece is the X-59 experimental aircraft, a piloted, single-engine jet being built by Lockheed Martin Skunk Works.

To understand the Quesst program is to appreciate that the X-59 is not a prototype for a commercial airliner but a single-use technology demonstrator that will gather data and demonstrate the efficacy of low boom flight technology. Its role is to validate the assumptions and theories around low boom technology to scientifically show that supersonic flight over land can be viable without disruption on the ground. If successful, it will mark a turning point in both environmental noise standards and the viability of high-speed air travel.

The aircraft is a data generation platform and its mission will support the total aerospace community, as well as advancing the current efforts of Boom Supersonic, Spike Aerospace and other potential operators of modern supersonic airliners. ICAO is planning to set new noise limits for supersonic aircraft

▼ **The first flight of the X-59 is due in late 2025.**
Lockheed Martin

▲ The X-59 was unveiled in January 2025. Lockheed Martin

and the issue of reducing the effects of sonic booms to communities below will determine if flights over land will be permitted.

NASA's Quesst

The X-59 is designed specifically to demonstrate low boom flight technology that reshapes the shockwaves produced during supersonic travel to generate a sonic thump, rather than the boom that led to the banning of supersonic flights over land. NASA will collect data that could make commercial supersonic flight over land possible, dramatically reducing travel time in the US or globally.

The Quesst mission has two main goals. First, to design and build NASA's X-59 research aircraft with technology that reduces the loudness of a sonic boom to a gentle thump for people on the ground. Second, to fly the X-59 over multiple US communities to gather data and public responses to the sound generated during supersonic flight and deliver that data set to national and international regulators. This is key because new entrants have differing views on how to mitigate sonic boom noise and the NASA programme will give regulators what they need to set defined limits about what is acceptable.

Quesst is organised within two of NASA's aeronautics programs: the Advanced Air Vehicles Program and the Integrated Aviation Systems Program. It is managed by a systems project office whose members span both projects and all four of the agency's aeronautical research field centres located in the US states of Virginia and Ohio, with two in California. The mission is comprised of three phases: building the X-59 aircraft, testing it in the air, then flying it over communities to survey what residents hear. The aircraft has been built and, by July 2025, it was finalising ground testing in preparation for its first flight later in the year.

Building X-59

In 2016, NASA launched the Quesst project and awarded a preliminary design contract to Skunk Works, the official nickname for Lockheed Martin's Advanced Development Projects. The project is part of NASA's New Aviation Horizons initiative that aims to accelerate development of environmentally responsible and efficient aircraft technologies.

After successful preliminary work, in April 2018 NASA awarded Lockheed Martin a $247.5m contract to build the full scale X-59 Quesst aircraft. Awarding that contract marked the transition from design to construction and the formal start of the X-59 programme, the centrepiece of NASA's aeronautical research strategy. The aircraft was built at the Skunk Works facility in Palmdale, California, with major components like the fuselage, wing and empennage produced separately and integrated in California. NASA's Armstrong Flight Research Center at Edwards Air

NASA X-59 EXPERIMENTAL

▲ The view pilots will have in the X-59.
Lockheed Martin

Force Base also played a critical role in preparing ground systems and infrastructure for flight testing.

The first flight was scheduled for 2021, but the programme was bedevilled by various delays and, by July 2025, the X-59 had only completed ground trials, with a first flight not expected until later in the year.

The critical design review was completed in 2020 and final assembly in 2022, although the aircraft was not unveiled until January 12, 2024, during a ceremony hosted by NASA deputy administrator Pam Melroy and other senior officials from Skunk Works. Melroy said: "This is a major accomplishment made possible only through the hard work and ingenuity from NASA and the entire X-59 team. In just a few short years we've gone from an ambitious concept to reality. NASA's X-59 will help change the way we travel, bringing us closer together in much less time."

At the unveiling, Bob Pearce, associate administrator for aeronautics research at NASA Headquarters in Washington, said it was thrilling to consider the level of ambition behind Quesst and its potential benefits. Pointing to the widespread benefit of the project, Pearce said: "NASA will share the data and technology we generate

▼ Engine runs provide a hint of the GE F414's power.
Lockheed Martin

www.key.aero 31

NASA X-59 EXPERIMENTAL

▶ The gleaming X-59 is helping return supersonic flight. NASA

▼ A single General Electric F414-GE-100 turbofan powers the X-59. Lockheed Martin

from this one-of-a-kind mission with regulators and with industry. By demonstrating the possibility of quiet commercial supersonic travel over land, we seek to open new commercial markets for US companies and benefit travellers around the world."

JAXA's connection

In July 2025, researchers from NASA and the Japanese Aerospace Exploration Agency (JAXA) tested a scale model of the X-59 experimental aircraft in a supersonic wind tunnel to assess the noise audible underneath the aircraft. Performed in Chofu, Japan, the wind tunnel testing was an important milestone in the search for technology to reduce the impact of sonic booms on communities and facilitate supersonic flight over land.

The JAXA wind tunnel measured 3ft long by 3ft wide, whereas the X-59 is 99.7ft long with a wingspan of 29.7ft. To perform the wind tunnel testing, researchers used a model scaled to just 1.62% of the actual aircraft aand measured 19 inches from nose to tail. They then exposed the model to conditions mimicking the X-59's planned supersonic cruising speed of Mach 1.4.

The testing performed at JAXA provided NASA researchers with critical data they could compare with their predictions derived through computational fluid dynamics modelling, including a clearer picture of how air would flow around the X-59 in flight. These marked the third round of wind tunnel testing following previous tests at JAXA and NASA's Glenn Research Center in Ohio, USA.

Shockwaves from traditional supersonic aircraft typically merge together, producing a loud sonic boom, but the X-59's unique design works to keep shock waves from merging, which will result in a quieter sonic thump. The wind tunnel testing provided researchers with data that helped them understand the noise level that would be created by the shockwaves the X-59 produced at supersonic speeds, to deliver design tools and technology for quiet supersonic airliners.

X-59 capabilities

The X-59 is expected to fly at Mach 1.4 at 55,000ft and its design, shaping and technologies will allow the aircraft to achieve these speeds while generating a quieter sonic thump. Perhaps its most distinctive feature was the long and thin tapered nose that accounts for almost a third of the aircraft's length. This spear-like nose breaks up the shockwaves coming off the aircraft. Another feature that might surprise many pilots and aviation engineers is that the cockpit on the X-59 is located almost halfway down the length of the aircraft and does not have a forward facing window. Instead, the Quesst team developed the eXternal Vision System (XVS), a series of high-resolution cameras and 4K monitors in the cockpit to give pilots a clear forward view on the ground and in the air. These unconventional adaptions were necessary to maintain the uninterrupted contour of the nose that is so important for supressing the sonic boom.

To maintain the X-59's sleek shape, NASA and Skunk Works opted for a canard controlled configuration where small, forward mounted control surfaces help manage the aircraft's stability

▲ The X-59 on the ramp and ready to go. Lockheed Martin

◀ Computer simulations provide valuable insights. NASA

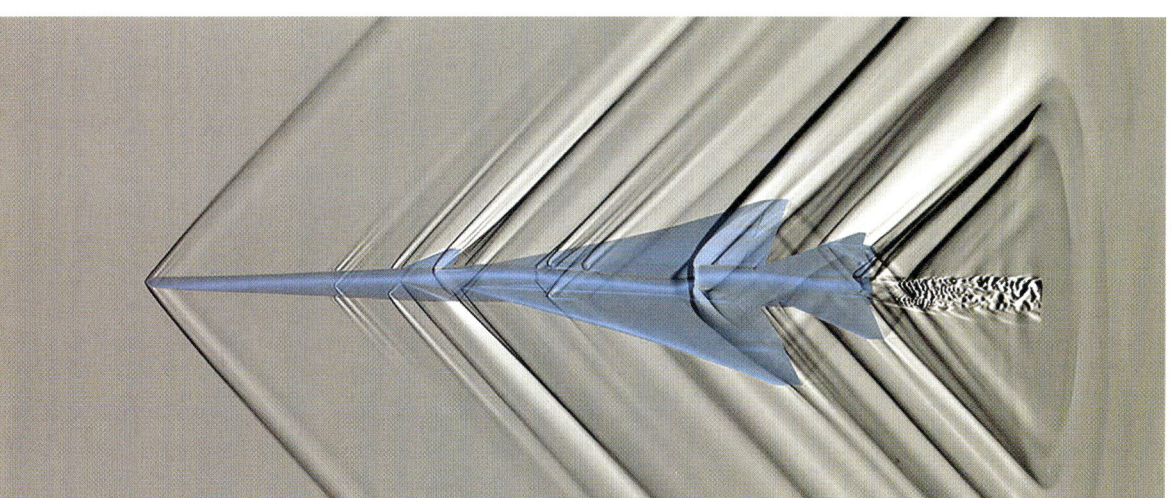

and control without disrupting the shockwave pattern. The wings are low-mounted and delta-shaped, while the engine is mounted on the top of the fuselage to reduce noise propagating downward. The Quesst team also mounted the engine on the top of the aircraft to give the X-59 a smooth underside and keep shock waves from merging behind the aircraft.

Super Hornet power

The X-59's power comes from a single General Electric F414-GE-100 turbofan that is a proven performer from its service on aircraft such as the F/A-18 Super Hornet. The engine delivers 22,000lb of thrust that will allow the X-59 to cruise at Mach 1.4 at altitudes of 55,000ft.

In October 2024, NASA started tests on the engine that would allow the X-59 team to verify the aircraft's systems were working together while powered by its own engine, whereas in previous tests the X-59 had used external power sources. The engine run tests set

◀ Schlieren photography assists engineers to understand sonic booms. NASA

NASA X-59 EXPERIMENTAL

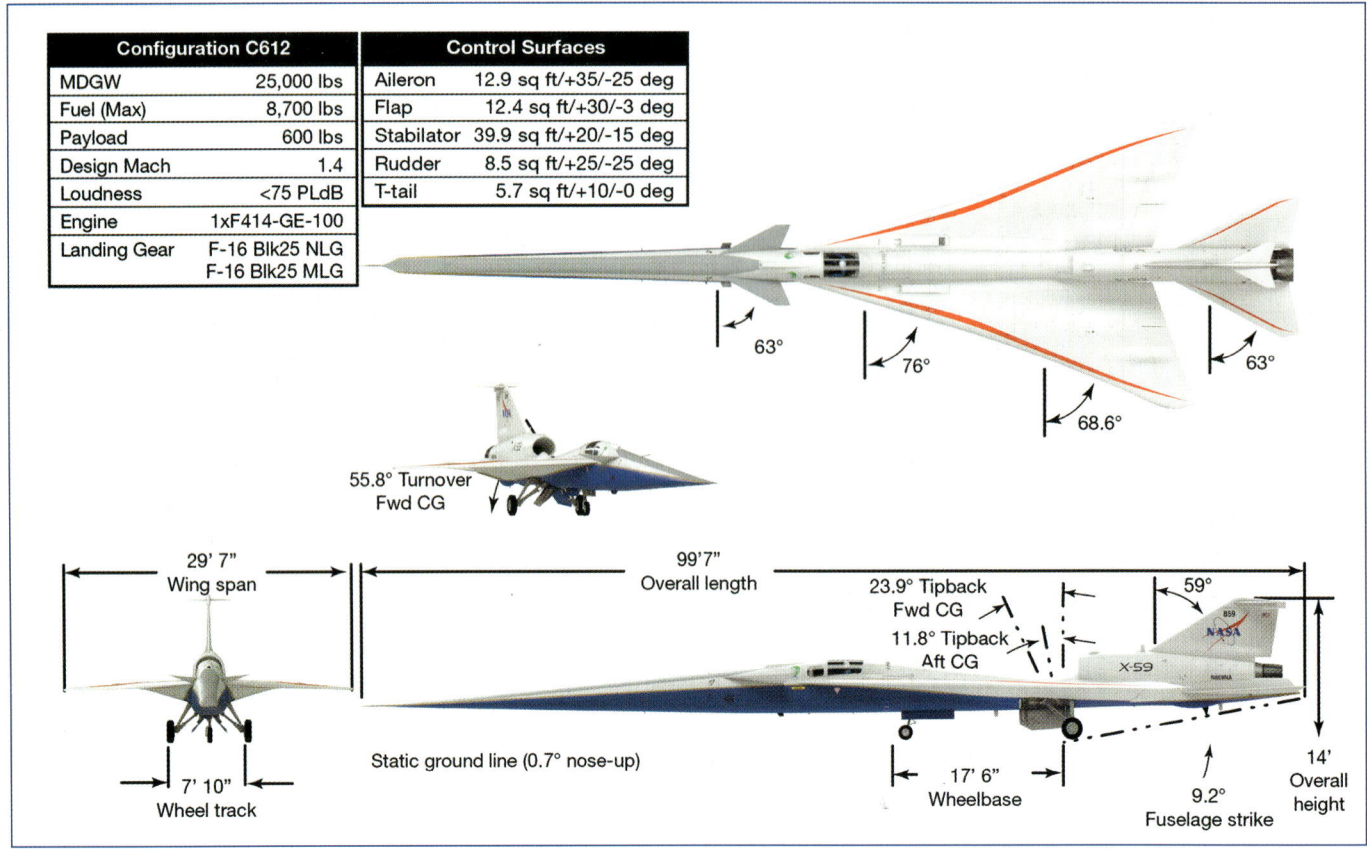

Configuration C612	
MDGW	25,000 lbs
Fuel (Max)	8,700 lbs
Payload	600 lbs
Design Mach	1.4
Loudness	<75 PLdB
Engine	1xF414-GE-100
Landing Gear	F-16 Blk25 NLG
	F-16 Blk25 MLG

Control Surfaces	
Aileron	12.9 sq ft/+35/-25 deg
Flap	12.4 sq ft/+30/-3 deg
Stabilator	39.9 sq ft/+20/-15 deg
Rudder	8.5 sq ft/+25/-25 deg
T-tail	5.7 sq ft/+10/-0 deg

▲ An overview of the X-59 capabilities. NASA

the stage for the next phase of the aircraft's progress toward its first flight in late 2025.

The tests were run in phases and in the first phase the engine was rotated at a relatively low speed without ignition to check for leaks and ensure that all systems were communicating as expected. The aircraft was then fuelled and the engineers began testing the engine at low power, with the goal of verifying that the aircraft systems operated without anomalies or leaks. NASA's chief engineer, Jay Brandon, explained that the first phase of engine testing was really a warm-up to make sure that everything looked good prior to moving to the actual first start and running the engine: "That took the engine out of the preservation mode that it had been in since installation on the aircraft. It was the first check to see that it was operating properly and that all systems it impacted – hydraulics, electrical system, environmental control systems, etc – seemed to be working."

Engine run tests are a routine part of the integrated ground tests needed to ensure safe flight and the X-59 team will continue progressing through these critical ground tests and address any technical issues found on this one-of-a-kind experimental aircraft. During these later phases the team will test the aircraft at high power with rapid throttle

▶ The X-59 will commence flight testing in 2025. NASA

NASA X-59 EXPERIMENTAL

◀ The tapered nose is key to quiet supersonic flight.
NASA

changes, followed by simulating the actual conditions of flight.

Ground testing

In July 2025, NASA commenced taxi tests, marking the first time the X-59 had moved under its own power. This was a significant milestone in the supersonic aircraft's development as taxiing represented the last series of ground checks before the X-59 is cleared for its first flight.

On July 10, 2025, at US Air Force Plant 42 in Palmdale, California, NASA test pilot Nils Larson and the X-59 team, comprised of NASA and Lockheed Martin personnel, completed the aircraft's first low speed taxi test. Engineers and flight crews monitored how the aircraft handled as it moved across the runway and worked to validate critical systems such as steering and braking. The next stage will see the X-59 gradually increase its speed, leading up to a high-speed taxi test that will take the aircraft just short of the point where it would take off.

Ground checks such as these help ensure the aircraft's stability and control across a range of conditions and give engineers and flight crews high levels of confidence that all systems are functioning as expected before the aircraft leaves the ground.

F-15s watching on

In May 2025, two NASA F-15 research jets made a series of flights to validate tools designed to measure and record the shockwaves that would be produced by the X-59. Carrying the recording tools, the F-15s flew faster than the speed of sound to match the conditions the X-59 was expected to fly in operations, such as when it performs supersonic flights over US communities.

The NASA team validated three key tools: a shockwave measuring device called a near-field shock sensing probe, a guidance capability known as an airborne location integrating geospatial navigation system and a Schlieren airborne photography system that will allow the capture of images that render visible the density changes in air caused by the X-59.

The F-15s completed a series of flight tests using two aircraft simultaneously. They flew in formation carrying near-field shock sensing probes and collected data from one another to test the probes and validate the tools under real-world conditions and help confirm how shockwaves form and evolve during flight.

For the Quesst mission, the F-15D led data gathering efforts using the onboard probe, while F-15B served as a back-up. The F-15 flew behind the X-59 and the probe helped measure small pressure changes caused by the shockwaves and validated predictions made some time ago when the aircraft's design was first created.

▼ The power of the GE engine glows brightly.
Lockheed Martin

NASA X-59 EXPERIMENTAL

▶ X-59's engine is fully prepared for its first flight.
Lockheed Martin

▼ Lockheed Martin Skunk Works is building the X-59.
Lockheed Martin

the quality of the data. To ensure the two aircraft were precisely positioned during the test flights, pilots used a NASA-developed software tool called the Airborne Location Integrating Geospatial Navigation System (ALIGNS). NASA researcher Troy Robillos, who led the development of the tool, explained: "ALIGNS acts as a guidance system for pilots. It shows them where to position the aircraft to either probe a shockwave at a specific point or to get into the correct geometry for Schlieren photography."

The ALIGNS software runs on tablet computers and uses GPS data from both aircraft to calculate when the aircraft are in position for probing and to capture the Schlieren imagery by giving pilots real time instructions enabling them to achieve precise positioning.

Schlieren photography provided Quesst researchers with crucial information. While computer simulations and wind tunnel tests that predict airflow are helpful, Schlieren imagery shows real-world airflow, especially in tricky zones like the engine and air inlet. The system utilises a handheld high-speed camera with a telescopic lens that captures hundreds of frames per second and visualises changes in air density, but only works if it can use the sun as a backdrop. The most challenging part is aligning two fast moving aircraft against the backdrop of the sun because the photographer must capture the aircraft flying across the centre of the sun and even the slightest shift can affect the shot and reduce

▶ Assembling the X-59 cockpit at Skunk Works.
Lockheed Martin

NASA X-59 EXPERIMENTAL

First flight

By August 2025, NASA had not yet confirmed a date for the first flight of the X-59, but with ground testing complete it is likely the aircraft will take off for the first time before the end of the year. NASA said the Quesst team would conduct several of the X-59 flight tests at Skunk Works before transferring the aircraft to its base of operations at Armstrong Flight Research Center and Edwards Air Force Base in California for validation. After NASA completes flight tests it will fly the aircraft over several selected cities across the US, collecting input about the sound the X-59 generates and how people perceive it.

Speaking about the progress to date, John Clark, vice president and general manager of the X-59 programme at Skunk Works, said: "Across both teams, talented, dedicated and passionate scientists, engineers and production artisans have collaborated to develop and produce this aircraft. We're honoured to be a part of this journey to shape the future of supersonic travel over land alongside NASA and our suppliers."

The other highly interested onlookers will be the teams at Boom Supersonic, Spike Aerospace, COMAC and others who are following a different path to eliminate or quieten the sonic boom. Organisations like NASA and JAXA are paving the way for new entrants globally to know exactly what they need to produce and the rules that govern where their new aircraft can travel at supersonic speeds. When ICAO presents its findings and recommendations there will be a level playing field for all to engage on with whatever technology they develop so long as it meets the standards for noise levels on the ground.

▲ With ground testing complete, the X-59 is primed for first flight.
Lockheed Martin

◀ NASA operates a wide variety of aircraft including F-15s. NASA

▼ NASA F-15s are a vital part of the X-59 Quesst project. NASA

www.key.aero 37

SUBSCRIBE TODAY
TO YOUR FAVOURITE MAGAZINE!

Combat Aircraft Journal is renowned for being America's best selling military aviation magazine.

Airliner World magazine is the biggest-selling commercial aviation magazine in the world.

SIMPLY SCAN THE QR CODE OF YOUR FAVOURITE MAGAZINE AND SUBSCRIBE TODAY!

Order today from our online shop
shop.keypublishing.com
Call +44 (0)1780 480404 *(Mon to Fri 9am - 5.30pm GMT)*

THE DESTINATION FOR
AVIATION ENTHUSIASTS
Visit us today and discover all our publications

Aviation News is renowned for providing the best coverage of every branch of aviation.

Airforces Monthly is devoted to modern military aircraft and their air arms.

and subscribe to your favourite magazine...
/collections/subscriptions

*Free 2nd class P&P on BFPO orders. Overseas charges apply.

▲ XB-1 is the first supersonic jet built with airliner technology. Boom

Boom Time

For more than 20 years Boom Supersonic has been making steady progress towards reviving global supersonic passenger transport.

Boom Supersonic started off in a typically entrepreneurial fashion with Blake Scholl founding the company in a basement in Denver, Colorado, in 2014. Since then, Scholl and the company have steadily progressed to the point where today they have designed, built and flown the world's first independently developed supersonic jet, which is also the first civil supersonic aircraft made in America.

Back in 2014, Concorde was long retired and the idea of commercial supersonic travel was fast fading into the annals of history, considered by many as too loud, too expensive and unsuited to a world increasingly aware of sustainability issues in aviation. As a frequent long-haul traveller, Scholl had a vision of building a Mach 1.7 aircraft capable of flying from New York to London in about 3.5 hours.

In 2016, Boom unveiled its plans by opening its first headquarters in Denver, with a team of experienced engineers and aerospace experts who had cut their teeth at Boeing, NASA, Lockheed Martin and SpaceX. Significantly, Scholl adopted Silicon Valley-style start-up principles – move fast, iterate and test – to get the Boom ball rolling, which first focused on building the XB-1 prototype.

Moving fast is one thing, but Scholl ensured that Boom adopted a phased development approach, starting with the XB-1 to validate the aircraft design and test critical technologies that would be needed for its Overture supersonic passenger airliner. In 2016, Boom also released a mock-up of the Overture and quickly gained solid interest from airlines including Virgin Atlantic and Japan Airlines.

Early interest

In December 2017, Boom announced it had formed a strategic partnership with Japan Airlines (JAL) to deliver supersonic passenger travel. For its part, JAL invested $10 million in Boom and said it would collaborate with Boom to refine the aircraft design and help define the supersonic travel passenger experience.

The partnership also gave JAL the option to purchase up to 20 Boom aircraft through a pre-order agreement, placing the airline in prime position to become one of the first in the world to offer supersonic passenger travel when Overture entered service. Scholl said the two had been working behind the scenes for more than a year and that JAL offered decades of practical knowledge and wisdom. He said: "We're thrilled

▲ Boom founder Blake Scholl wants to bring the benefits of supersonic flight to all. Boom to be working with JAL to develop a reliable, easily maintained aircraft, providing revolutionary speed to passengers. Our goal is to develop an airliner that will be a great addition to any international airline fleet."

Environmentally friendly

Boom unveiled the XB-1 demonstrator in 2020 as the world's first fully carbon neutral aircraft programme through the use of sustainable aviation fuels and carbon offsetting. Concorde had been many things, but it was never considered a model for sustainable or environmentally friendly aviation. Boom is determined to be on the right side of that ledger.

Boom said it was ushering in a new supersonic era where environmental considerations were essential throughout the aircraft's design, testing and flying. It integrated sustainability considerations into major decisions and Boom's commitment covered all ground and flight testing performed over the lifetime of the XB-1 programme.

In January 2019, Boom had successfully conducted a series of ground tests running XB-1 engines using a blend of more than 80% sustainable aviation fuel (SAF), giving Boom the confidence it needed to use SAF in future ground and test flights. In June 2019, Boom announced a partnership with sustainable alternative fuel innovator Prometheus Fuels, which removes CO_2 from the air and uses renewable energy to turn it into zero net-carbon jet fuel.

XB-1 takes off

By August 2023, Boom had completed several key milestones on the XB-1's pathway into service and had moved the demonstrator

BOOM TIME

▲ **Test pilot Tristan Brandenburg preparing for an XB-1 supersonic flight test.** Boom

from the company's hangar in Colorado to the Mojave Air and Space Port in California. The aircraft had already begun ground testing in Colorado in 2020, where tests of all its internal systems were performed, along with upgraded landing gear and supersonic engine intakes, both of which increased safety and performance.

The XB-1 has a carbon composite and titanium fuselage and is powered by three GE J85 jet engines that produce a combined maximum thrust of 12,300lb of force. After a detailed aircraft inspection, Boom received an experimental airworthiness certificate from the FAA in 2023, as well as letters of authorisation to allow chief test pilot Bill 'Doc' Shoemaker and test pilot Tristan 'Geppetto' Brandenburg to operate the XB-1.

The big day arrived on March 22, 2024, when the XB-1 took off from Mojave on its inaugural flight. The aircraft was flown by Bill Shoemaker, while Tristan Brandenburg manned the T-38 chase aircraft monitoring the flight. This took place in the same airspace that legendary pilot Chuck Yeager had broken the sound barrier in 1947, as well as testing of the North American X-15 and Lockheed SR-71 Blackbird. The XB-1 met all of its objectives including safely and successfully achieving an altitude of 7,120ft and speeds up to 238kt.

▶ **Overture wants to get FAA and EASA certification by the end of the decade.** Boom

▲ XB-1 test flights are operated from the Mojave Air and Space Port in California. Boom

Scholl commented: "Today, XB-1 took flight in the same hallowed airspace where the Bell X-1 first broke the sound barrier in 1947. I have been looking forward to this flight since founding Boom in 2014 and it marks the most significant milestone yet on our path to bring supersonic travel to passengers worldwide."

In the aftermath of the inaugural flight, the XB-1 completed a rigorous series of 11 test flights under increasingly challenging conditions to evaluate systems and aerodynamics. Using data from those test flights, the Boom team systematically expanded the flight envelope through subsonic, transonic and supersonic speeds

The XB-1 programme will form the foundation for the design and development of the Overture airliner and has validated some of its key technologies and innovations:

- **Augmented reality vision system** Two nose-mounted cameras, digitally augmented with attitude and flight path indications, feed a high-resolution pilot display enabling excellent runway visibility. This system enables improved aerodynamic efficiency without the weight and complexity of a moveable nose.
- **Digitally optimised aerodynamics** Engineers used computational fluid dynamics simulations to explore thousands of designs for XB-1. The result is an optimised design that combines safe and stable operation at take-off and landing with efficiency at supersonic speeds.
- **Carbon fibre composites** XB-1 is almost entirely made from carbon fibre composite materials, enabling it to realise a sophisticated aerodynamic design in a strong, lightweight structure.
- **Supersonic intakes** XB-1's engine intakes slow supersonic air to subsonic speeds, efficiently converting kinetic energy into pressure energy, allowing conventional jet engines to power XB-1 from take-off through supersonic flight.

Following the successful maiden flight Boom quickly turned its attention to the next milestone of breaking the sound barrier and beginning to scale up the technology on XB-1 for the Overture airliner.

Breaking the barrier

On January 28, 2025, XB-1 accelerated through the sound barrier for the first time, reaching Mach 1.222 at 35,290ft. This was the first time an independently developed jet had achieved this milestone, with previous supersonic aircraft being the work

BOOM TIME

▶ Boom's chief test pilot Tristan Brandenburg in the XB-1. Boom

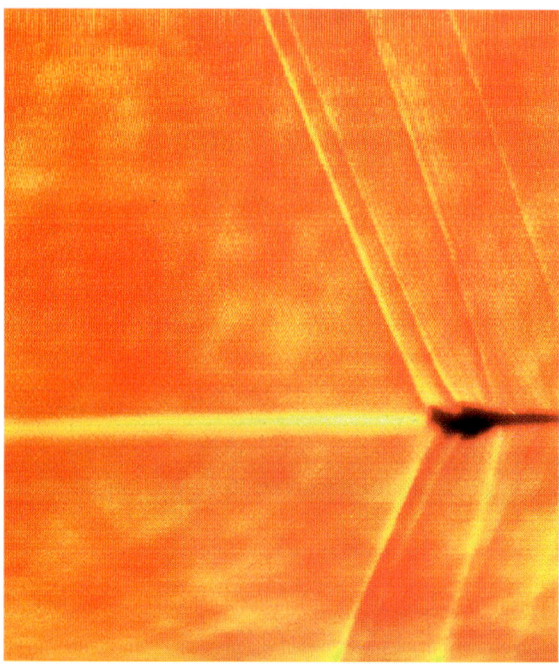

▼ Schlieren image of the XB-1 breaking the sound barrier. Boom

of nation states and developed by governments and militaries.

Scholl commented: "XB-1's supersonic flight demonstrates that the technology for passenger supersonic flight has arrived. A small band of talented and dedicated engineers has accomplished what previously took governments and billions of dollars. Next, we are scaling up the technology on the XB-1 for the Overture supersonic airliner (and) our ultimate goal is to bring the benefits of supersonic flight to everyone."

Depending on the airline cabin configuration, Overture will carry 64-80 passengers at Mach 1.7, which is around twice the speed of today's subsonic airliners. Former chief Concorde pilot for British Airways, Mike Bannister said he had been waiting more than 20 years for the return of supersonic speeds: "XB-1's historic flight is a major landmark towards my dreams being realised. When I last flew Concorde in 2003, I knew this day would come. Boom is well on its way towards making sustainable supersonic flight a reality aboard Overture, my number one choice as a successor to Concorde. Congratulations to all at Boom and especially its pilot, Tristan 'Geppetto' Brandenburg. Having been Concorde's chief pilot, I envy his role in this exciting return towards mainstream supersonic flight."

On February 10, the XB-1 made its second supersonic flight, reaching new heights of 36,514ft and a top speed of Mach 1.8. During the 41-minute flight, the XB-1 broke the sound barrier three times, which Boom said was achieved without creating sonic booms that could be heard on the ground.

The objective of the flight was to continue to assess aircraft performance above Mach 1, including capturing acoustic sonic boom data and Schlieren photography to visualise the shock waves resulting from XB-1 pushing through the air at supersonic speeds. It marked the conclusion of XB-1's flight test program and the aircraft returned to its home base in Colorado.

Schlieren imaging

Schlieren imaging is a powerful optical technique used in aerospace engineering and aviation to visualise changes in air density, particularly those caused by shock waves, heat gradients and pressure differentials. It is perfectly

▶ Boom XB-1 first broke the sound barrier on January 28, 2025. Boom

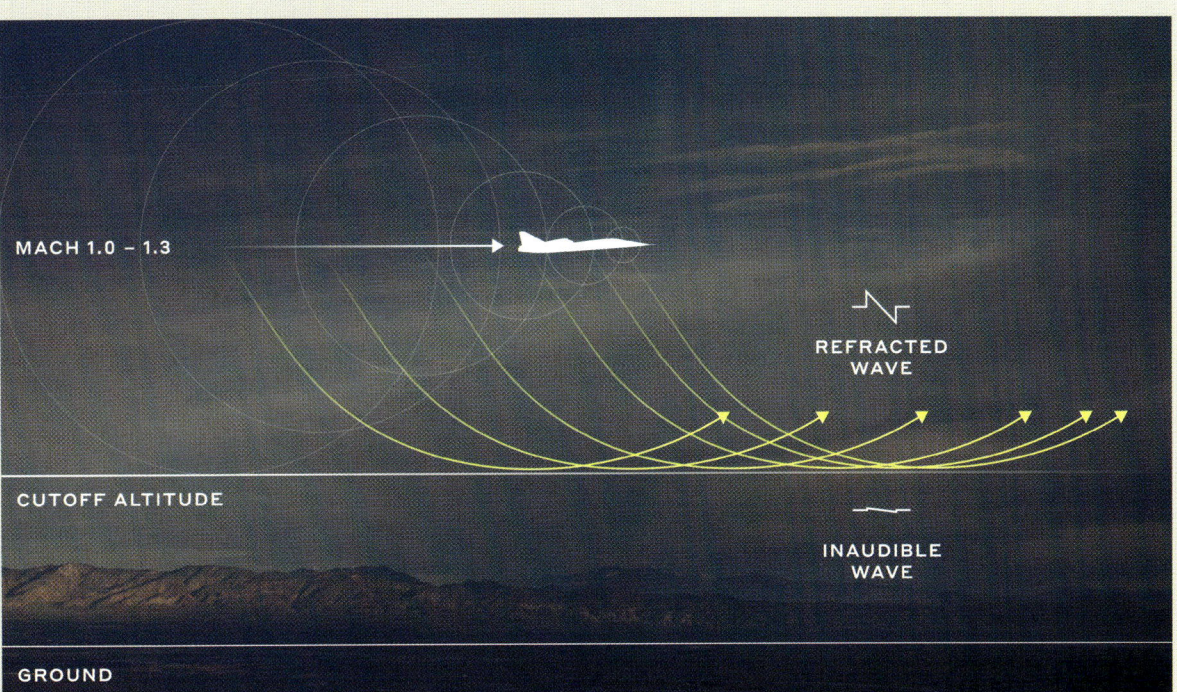

▸ Boomless cruise uses Mach cut-off to avoid sonic booms reaching the ground. Boom

suited for studying supersonic and hypersonic flight where airflows become complex and highly compressed.

Invented by German physicist August Toepler in 1864, it works by detecting tiny changes in the refractive index of air. When air density changes due to speed, pressure or heat, it bends light slightly and Schlieren systems make these bends visible by shining light through the airflow and capturing distortions with sensitive cameras or mirrors.

Airborne Schlieren is being used by NASA to study the shockwaves from the X-59 supersonic demonstrator by taking images of the aircraft in flight using high altitude cameras and the sun as a light source. In a clear sign of its methodical development process, Boom partnered with NASA to utilise the technology on the second XB-1 supersonic test flight on February 10, 2025.

◂ Boom has assembled an impressive group of partners and suppliers for Overture. Boom

Taking Schlieren images requires ideal conditions and timing, as well as exceptional flying. In Boom's case, that demand fell on the shoulders of Tristan Brandenberg. He was able to position XB-1 at an exact time in a precise location over the Mojave Desert to enable NASA to photograph it flying in front of the sun, documenting the changing air density around the aircraft at speeds exceeding Mach 1. The Boom team used waypoints computed by NASA to develop avionics software to guide Brandenberg to the specific point

▾ The XB-1 broke the sound barrier six times without audible sonic booms on the ground. Boom

BOOM TIME

▲ Political will in the US is shifting towards over land supersonic flights.
Boom

that XB-1 would have to fly through in order to eclipse the sun. NASA captured the imagery using ground telescopes with special filters.

Commenting on the spectacular outcome, Scholl said: "Schlieren imaging makes the invisible visible – the first American made civil supersonic jet breaking the sound barrier. Thanks to Tristan's exceptional flying and our partnership with NASA, we were able to capture this iconic image."

Boomless cruise

If Overture is to be a commercial success it will need to fly supersonic over land, the one part of the puzzle that Concorde was never able to achieve. There is growing political will, especially in the US, to lift the ban on such flights, but only if operators can reduce the sonic booms reaching the ground to stringent new levels.

During the second test flight, NASA teams collected data on XB-1's acoustic signature at a location on the flight route as the jet accelerated. Boom said its analysis found that no audible sonic boom reached the ground as the jet flew at

supersonic speeds, which Scholl said will "pave the way for coast-to-coast flights up to 50% faster."

To address the sonic boom noise issue, Boom is relying on well-established physics known as 'Mach cut-off', in which a sonic boom refracts in the atmosphere and never reaches the ground. The effect is achieved by breaking the sound barrier at a high enough altitude with exact speeds varying on atmospheric conditions. XB-1's second supersonic flight demonstrated this capability and, in total, the aircraft broke the sound barrier six times without making an audible boom. Scholl said: "If we can control the way a sonic boom comes off the airplane – making it come off at a shallow angle while flying sufficiently high – we can actually make the sonic boom take a U-turn in the atmosphere before it ever hits the ground. We call this capability the boomless cruise. It uses real-time weather data plus advanced computers and powerful engines to fly reliably in the Mach cut-off condition."

On Overture, boomless cruise is enabled by the Symphony engines, which feature enhanced transonic performance compared to commercially derived engines, allowing Overture to efficiently transition to supersonic speeds at altitudes above 30,000ft. The advanced autopilot uses real-time

◂ **Blake Scholl opened the Boom Overture Superfactory in 2024.** Boom

◂ **Overture launch customers include United Airlines, American Airlines and Japan Airlines.** Boom

◂ **Boom's XB-1 preparing for its second supersonic flight test.** Boom

BOOM TIME

▲ The Overture Superfactory will incorporate an aircraft delivery centre. Boom

▼ By mid-2025, Boom had secured commitments for 130 Overtures. Boom

weather conditions and software algorithms to automatically select the optimal speed for boomless cruise.

So why was this technology not used on Concorde or other supersonic aircraft? Boom has noted: "Theoretically, any supersonic aircraft at the right speed and altitude can achieve Mach cut-off. However, Concorde lacked the engine efficiency to fly in Mach cut-off without afterburners, making the range impractical. Additionally, in Concorde's era, the computing power to calculate appropriate speeds and altitudes was not available."

One major difference between Concorde and Overture is that the latter takes off without afterburners, giving it the same landing and take-off noise footprint as today's subsonic long-haul aircraft. Overture will also incorporate a variable noise reduction system that automatically manages thrust to reduce noise at take-off and Symphony will be considerably quieter than Concorde's turbojets with afterburners.

Facilities

Boom is also well advanced in developing the airframe and engine facilities it needs to support the Overture programme. In June 2024, it celebrated the construction of the Overture Superfactory at the Piedmont Triad International Airport in Greensboro, North Carolina. This first assembly line has the capacity to produce 33 Overture aircraft per year, valued at more than $6b, and Boom has plans to build an additional assembly line and scale up to 66 aircraft annually. The Overture Superfactory campus will also include a delivery centre where airlines will take delivery of their supersonic airliners.

Scholl said: "Construction of the Overture Superfactory represents a major milestone toward ensuring continued US leadership in aerospace manufacturing. Supersonic flight will transform air travel and Overture provides aninnovative alternative for airlines across the globe."

In April 2025, Boom announced it had selected a site at the Colorado Air and Space Port to conduct engine tests for its Symphony propulsion system. Developed by Boom, Symphony is a medium bypass turbofan optimised for supersonic flight. The Colorado site was previously used for hypersonic engine development and Boom is investing up to $5m to support 2025's test of Symphony's core, the high-pressure spool of the engine consisting of the compressor, combustor and turbine. Boom expects to expand the site in 2026

BOOM TIME

◀ Boom has integrated sustainability considerations into all its major decisions. Boom

◀ Overture will carry 64-80 passengers at Mach 1.7, around twice the speed of today's subsonic airliners. Boom

Latecoere, Leonardo, Safran Landing Systems, Universal Avionics and the USAF.

In August 2025, Boom said it plans to start production of the Overture in 2026, with a goal of rolling out the first aircraft in three years and be conducting flight tests in four. Overture also has a target of achieving FAA and EASA certification to carry passengers by the end of the decade. Based on its progress to date, there seems no reason to doubt that Brad Scholl and Boom will achieve that goal and transform commercial aviation in the most profound way.

to facilitate full scale testing of the entire Symphony turbofan prototype.

Symphony passed a key technical review in March 2025, enabling manufacturing launch. Multiple parts have already entered the manufacturing phase and engine assembly will start in late 2025. When the programme moves into production, Boom will leverage the team and facilities of Standard Aero in Texas to build engines at rate.

Where next?

As of April 2025, Boom advised that its order book stood at 130 aircraft, including pre-orders from American Airlines, United Airlines and Japan Airlines. It is also working with Northrop Grumman in developing government and defence applications for Overture. Boom has assembled a high-quality collection of partners and suppliers for the Overture project, including Aernnova, Aciturri, Collins Aerospace, Eaton, Honeywell,

▼ The first assembly line has the capacity to produce 33 Overtures per year. Boom

CHINA CONNECTION

China Connection

China wants to lead the global aviation industry and sees an opportunity to demonstrate its technological prowess with supersonic and hypersonic aircraft.

CHINA CONNECTION

Since its inception, commercial aviation has been driven by major US and European corporations and, in the last decade, it has been Airbus and Boeing who have set the pace with their new-generation aircraft. Both companies have had significant success in the Chinese market, but that cosy duopoly is now under threat as China boldly embarks on a mission to dominate the global aviation industry.

Driven by the government's Made in China 2025 and 2035 Vision programmes, China is openly challenging Western dominance in aviation and sees an opportunity to leapfrog current aviation technology by developing supersonic and hypersonic passenger aircraft. The long-term objective is to create a viable, competitive and sustainable supersonic transport aircraft capable of Mach 2-4 operating transcontinental services by the 2030s.

Organisations such as the Aviation Industry Corporation of China (AVIC), the Commercial Aircraft Corporation of China (COMAC) and the China Aerospace Science and Technology Corporation (CASC) have teamed up with research organisations like the Chinese Aeronautical Establishment and Chinese Academy of Sciences to develop a supersonic airliner. The work is not just about supersonic speeds, but also takes into account sustainability, environmental regulations and economic viability. To address those, the search is on for new aircraft materials, techniques to reduce sonic booms over land and more efficient propulsion systems.

Supersonic C949

COMAC has its work cut out for the next few decades as it has been entrusted to spearhead China's bid to lead the future of global aviation. The most talked about and promoted project is the family of commercial jet airliners intended to compete head on with Airbus and Boeing on a global basis. Aircraft like the in-service ARJ21 advanced regional jet and C919 narrowbody and the in-development C929 widebody will be significant within China's large domestic and international airlines and challenge the two main incumbents. They will also be viable options to many carriers in the Asia-Pacific region, one of the fastest growing and largest commercial aviation markets in the world.

The Chinese government is proud of its technological progress and embedded in its aviation leadership aspiration is showcasing on a global scale the innovation, talent and manufacturing skills of major aerospace companies such as COMAC. With such a challenging national goal it is unsurprising that China also sees supersonic air travel as another sector where global leadership is attainable.

In March 2025, COMAC quietly unveiled the C949 supersonic airliner not at an elaborate media function but as an academic paper in *Acta Aeronautica Sinica*, a Chinese semi-monthly journal that was first published in 1965. The journal is published by the editorial office of Acta Aeronautica et Astronautica Sinica, sponsored by Chinese Society of Aeronautics and Astronautics and Beihang University and superintended by China Association for Science and Technology. Its charter is to reflect the development level of science and technology in China's aerospace field, exchange new achievements in science and technology at home and abroad, accelerate academic progress and talent growth and promote the development of new theories and technologies. Its audience includes researchers of scientific research institutions in the field of aerospace technology, as well as teachers and graduate students of aerospace related specialties in colleges and universities.

In the paper, COMAC revealed the blueprint of a Mach 1.6 airliner that promised significantly greater range and a much quieter noise footprint than Concorde. The range increased from Concorde's 4,500 miles to 6,800 miles, while the sonic boom was slashed from 105PLdB to 83.9PLdB, which COMAC described as comparable to the noise level of a hairdryer. The designers claimed the noise reduction was achieved through advanced aerodynamic shaping, including a reverse-camber

◀ **COMAC is considering launching a new supersonic jet to follow Concorde.** Graham Bloomfield/Shutterstock

▼ **COMAC is studying a potential low boom supersonic airliner.** COMAC via Acta Aeronautica

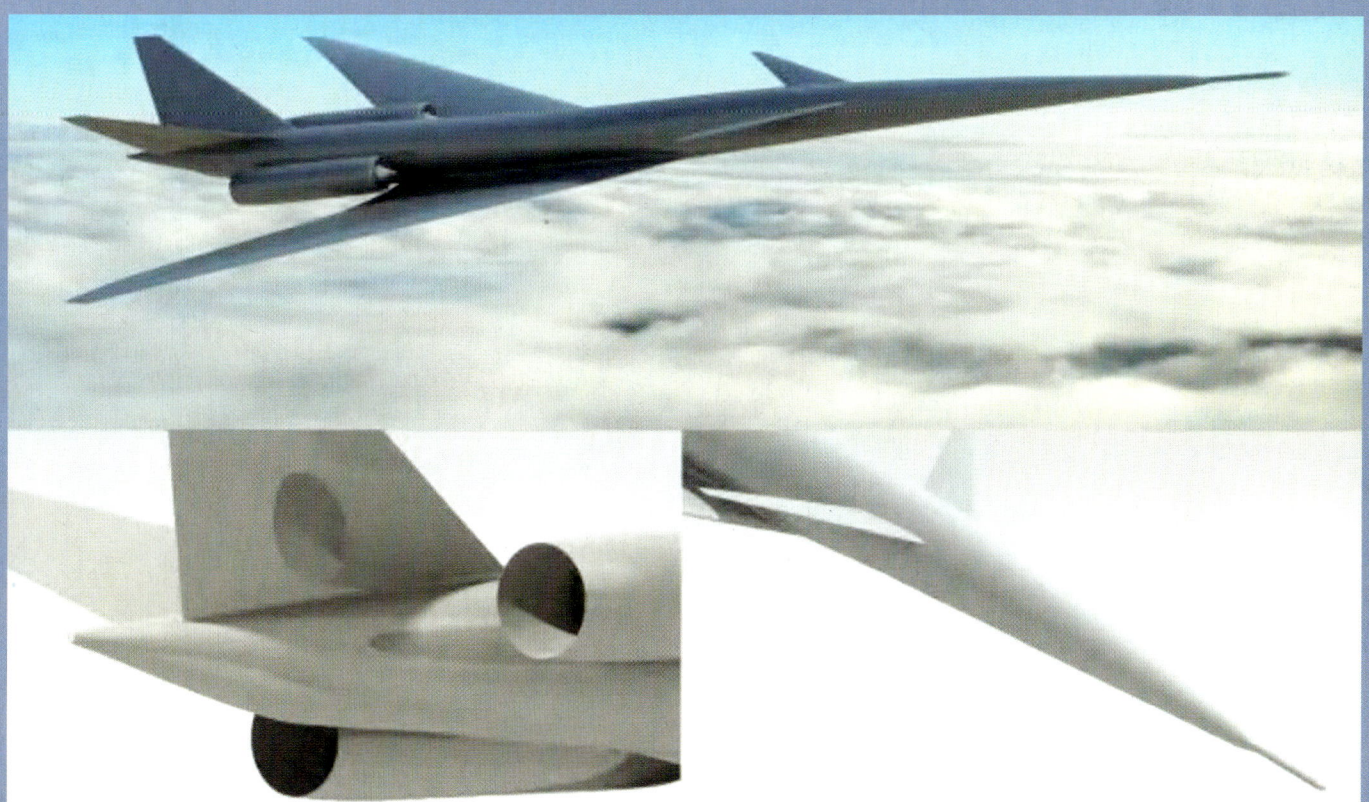

CHINA CONNECTION

▶ **The world's first COMAC C919** Coby Wayne/Shutterstock

fuselage and other low-boom optimisation techniques.

The major impediment to commercially viable supersonic passenger services is the ban on over land flying, although the cost of tickets and growing sustainability concerns are not to be discounted. It is not airframe or engine technology that is holding these flights back, so the first entrant to reduce the sonic boom to an acceptable level will take leadership.

In that regard, the research work using the NASA/Lockheed Martin X-59 experimental aircraft in late 2025 to assist ICAO in setting universal supersonic noise limits feeds directly into the development profiles for new manufacturers, such as COMAC, Boom Supersonic and Spike Aerospace. As to whether their aircraft will be as quiet as a hairdryer or a car door closing, only time will tell when actual flight testing is undertaken, with the great unknown being the earth's atmospheric conditions.

Apart from the ban on over land flying, Concorde's commercial success was hampered by its limited range, which restricted it mainly to transatlantic services. The extra range of the C949 would allow for routes connecting city pairs such as Beijing-Los Angeles or Shanghai-New York, adding more opportunities than those previously focused on Europe to US services.

C949 Numbers

COMAC has not announced a timeline for the first flight, certification or entry into service of the C949, although speculation is that it will be in 2049, which is the centenary of the People's Republic of China. As well as the patriotic significance of that anniversary it also leaves COMAC with more than two decades to develop and test the aircraft, which is an extraordinarily long timeline given China's resources and what is already known about supersonic flight.

While the March 2025 article in *Acta Aeronautica Sinica* was informative, from a technical perspective it posed more questions than answers and lacked a definitive set of specifications for the new aircraft. The journal did disclose the following key points:

▼ **China Southern Airlines COMAC ARJ21** Mario Hagen/Shutterstock

CHINA CONNECTION

▲ COMAC ARJ21 operated by Chengdu Airlines
Markus Manika/ Shutterstock

- The C949 is a long-range twin-engine supersonic airliner configured to carry 28-48 passengers in a business class configuration or up to 168 in a mixed layout.

- It will cruise at Mach 1.6, with flexibility for Mach 1.7 in Eco mode at high altitude

- It has a range of 6,800 miles, around 50% farther than Concorde
- It's powered by twin adaptive-cycle turbofan engines mounted on the rear fuselage
- It has fuel capacity of 42,000kg in seven distributed tanks with dynamic inflight centre of gravity management

- AI enhanced fly-by-wire systems will enable supersonic stability control
- It has a reverse camber mid-fuselage, elongated needle nose and engine bulges for shock dispersion
- The super-swept delta wings, canards and V-tail arrangement will suppress sonic boom intensity

These specifications are not official, but are derived from the published academic paper and media summaries of those disclosures. Given that the C949 project is at a very early conceptual stage it is understandable that any detailed manuals and specifications are proprietary and not yet publicly available.

COMAC

The Commercial Aircraft Corporation of China was founded in May 2008 to develop large civil aircraft and reduce China's reliance on foreign manufacturers such as Airbus and Boeing. It is owned by China's Assets Supervision and Administration Commission (SASAC) of the State Council and is a central part of China's long term goal to build an independent commercial aviation industry.

COMAC's slogan is 'Let China-made large aircraft fly in the blue sky' and the company was created using civil aviation design and manufacturing units from AVIC. The creation of COMAC was part of the much broader national strategy to promote indigenous innovation and high-tech development under the banner of Made in China.

When COMAC was formed it inherited the ARJ21 from AVIC, a 78-90 seat advanced regional jet

CHINA CONNECTION

▶ **China Southern has a fleet of COMAC C909 regional jets** Danny Ye/Shutterstock

powered by GE CF34 engines that performed its first flight in 2008. The ARJ21 development programme, later renamed the C909, suffered numerous delays related to design revisions, aerodynamic instability and certification challenges before finally entering commercial service in 2016 with Chinese carrier Chengdu Airlines.

Perhaps the most important aircraft COMAC will ever produce is the C919 narrowbody, which was announced in 2008 to compete directly with the Airbus A320 Family and Boeing 727 MAX airliners. The rapid growth of China's aviation sector has come off the back of Airbus and Boeing aircraft, but that reliance is already shifting, with major airlines such as China Eastern, Air China, China Southern and others placing orders for more than 1,200 C919s by 2025.

In 2024, the C919 carried its one millionth revenue passenger and operated its first route outside mainland China with flights between Shanghai and Hong Kong. The aircraft is awaiting certification by the US Federal Aviation Administration (FAA) and the European Aviation Safety Agency (EASA), but as China has mutual recognition of airworthiness certification with Singapore, Indonesia and Myanmar, the C919 can already operate in those nations.

By mid-2025, Airbus and Boeing had a combined backlog of more than 15,000 aircraft, which equated to around 13 years of production at 2024 annual delivery rates. The vast majority of those aircraft are A320neo Family or 737 MAX jets. With the Asia-Pacific region now the world's largest commercial aviation market, holding onto to market share in the face of competition from the locally built C919 is vital for Airbus and Boeing.

▼ **The COMAC team at the Farnborough Air Show in the UK** CSWFoto/Shutterstock

Flying Monkeys

There are many aerospace or entrepreneurial businesses stepping forward to release concepts for supersonic passenger transport, although thankfully not as many as there were in the early days of eVTOLs or unmanned aerial vehicles.

Sitting alongside the COMAC 949 programme, China is also developing much faster hypersonic transport concepts with dual military-civilian potential, such as the Yunxing (Space Transportation) hypersonic spaceplane. Space Transportation, also known as Lingkong Tianxing Technology, is a private project aiming to build aircraft capable of Mach 4-6 for both rapid point-to-point travel and space tourism.

In January 2025, the Lingkong Tianxing Technology Co of Beijing unveiled the prototype of its next-generation supersonic aircraft, the Cuantianhou (Soaring Monkey) from the Yunxing series. The prototype transport plane is designed to fly at Mach 4 and operate in near-space approximately 20-100 kilometres above Earth.

The company said it expects to operate the Cuantianhou's maiden flight in 2026 to test aerodynamics, power systems

CHINA CONNECTION

and heat resistant materials. This follows successful tests of the Yunxing prototype supersonic plane that achieved a flight speed of more than Mach 4 as the development entered the engineering phase. Speaking at a conference in Chengdu in January 2025, Lingkong Tianxing Technology's chief engineer, Deng Fan, said that the prototype of the Cuantianhou will be seven metres long and weigh 1.5 tonnes. It will use a streamlined aerodynamic design that can effectively reduce air drag and improve flight efficiency. The aircraft will be propelled by a state-of-the-art ramrotor detonation engine which combines a rotary detonation engine, rotor compressor and ramjet technology. The 1.9 metre Jindou-400S engine system will weigh 100kg and create thrust of at least 4,000 newtons.

Deng said: "Supersonic passenger aircraft are making significant breakthroughs and a strong comeback. There is great potential to provide more economical and reliable high-speed flight services in the coming years. Using this engine will allow Cuantianhou to fly at a top speed of Mach 4.2, which means the craft will be almost five times faster than a conventional jetliner."

The aircraft will be launched by a carrier rocket that will take it to an altitude of about 6,500ft, after which it will continue to fly on its own, although *China Daily* said that Deng declined to disclose how the aircraft would land. Near space encompasses portions of the stratosphere, mesosphere and lower thermosphere and is above the top altitudes of commercial airliners but below orbiting satellites.

The company said that if everything goes according to plan the prototype for a commercial supersonic airliner name Dasheng (Monkey King) is expected to conduct its maiden flight in 2030, which Deng claimed would revolutionise the global air transportation sector.

▼ A drawing of Space Transportation's Soaring Monkey under development in China Space Transportation

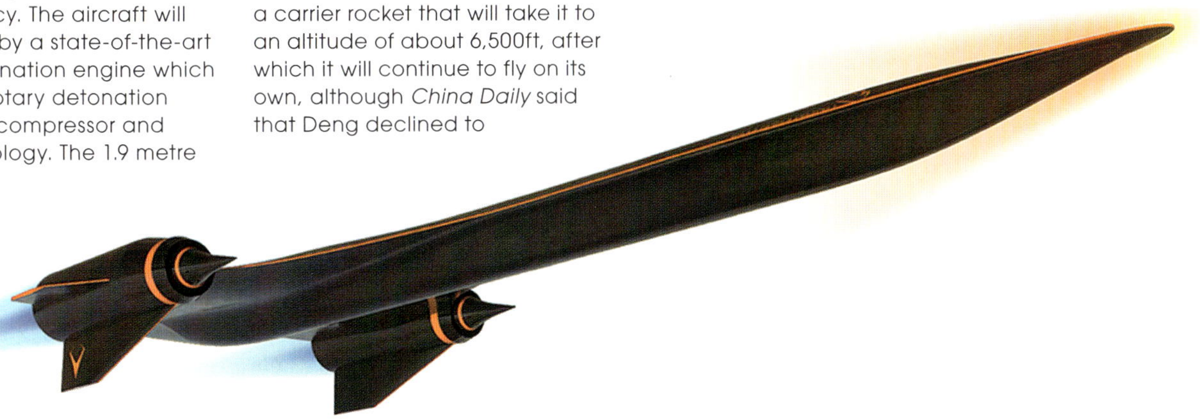

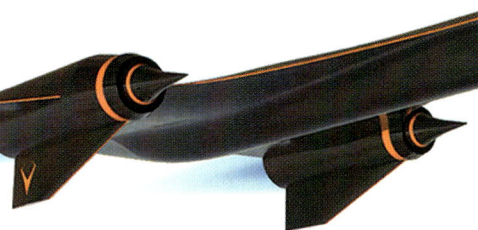

CHINA CONNECTION

▶ COMAC is now developing the widebody C929.
Michael Doran

CHINA CONNECTION

◀ COMAC aircraft are gaining market share in Southeast Asia. Minh Tuan Pham/Shutterstock

Supersonic, which is focused on supersonic commercial passenger transport for civilian use without the comfort of state-owned backing or financial support.

While it is generally true that advances in military aviation often filter into commercial aircraft programmes and vice versa, China's holistic approach of melding the two and then backing the programme to the hilt to achieve a nationally important strategic goal is far superior to relying on venture capitalists or wealthy entrepreneurs to revive supersonic passenger travel. This integration gives China a distinct competitive advantage in that research in one domain directly supports the other, with technologies such as advanced ramjets, rotating detonating engines, thermal protection systems, high strength composite materials and shockwave control being developed simultaneously for hypersonic weapons and for future passenger aircraft.

Overlaid on technology is a supportive regulatory environment and a central planning model that aligns research goals, advanced manufacturing needs, talent acquisition, educational needs and, most importantly, provides funding that supports long-term projects even when commercial returns are uncertain or not expected to deliver for decades to come.

Contrast China's approach with the uncertain regulatory and commercial environments in which organisations like Boom Supersonic, Spike Aerospace, Lockheed Martin and NASA operate to achieve the same goal and it seems the race for global aviation leadership is becoming a little one-sided.

Dual use aircraft

Space Transportation said it has plans to conduct full scale test flights of Cuantianhou by 2027, with a long term goal of connecting Beijing and New York in as little as two hours. This project is unique in that its use of sub-orbital or near-space flights transcends the line between commercial aviation and space travel.

It is also different to other projects in that it is combining supersonic and hypersonic technologies that are underpinned by China's political and national goals and the fusion of military and civil policy and research. That's in contrast to Boom

▼ COMAC is a regular exhibitor at leading global air shows. Jordan Tan/Shutterstock

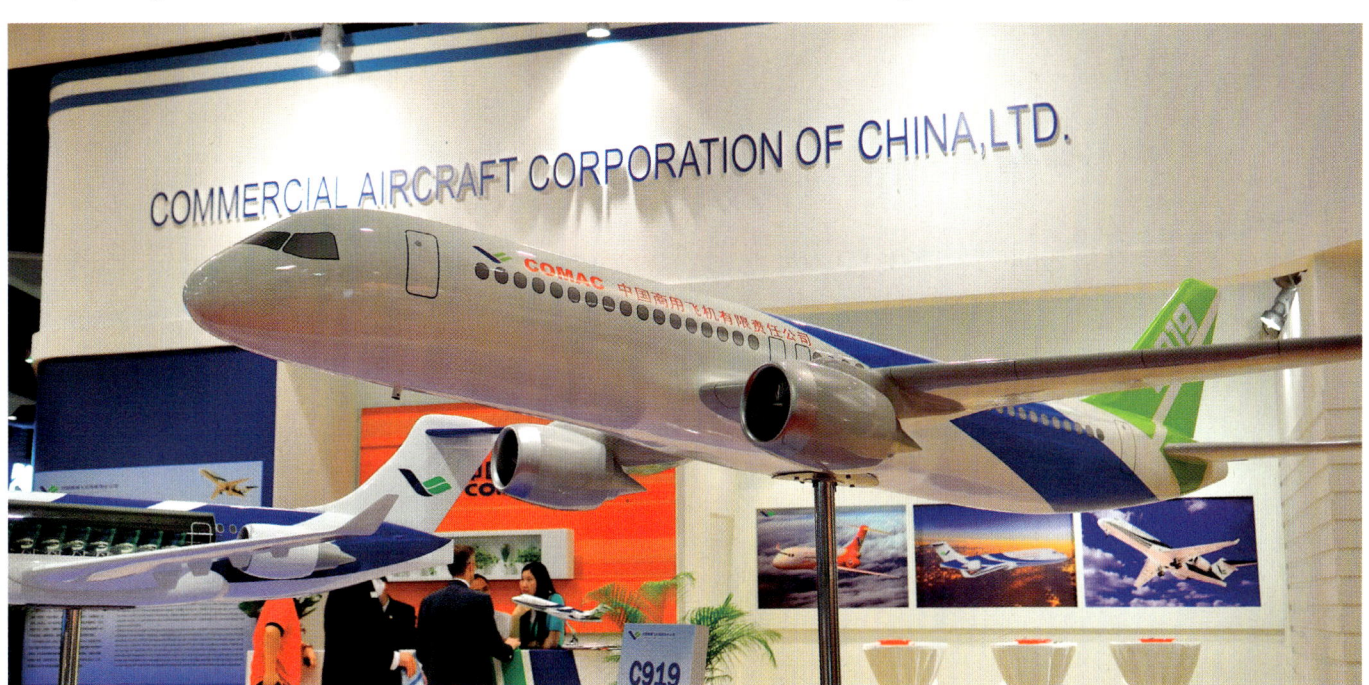

Hits and Misses

While several hopefuls are announcing new supersonic passenger transport projects, the same issues that stunted Concorde's growth have yet to be resolved.

More than 20 years since Concorde's last commercial flight there is a real possibility that supersonic passenger transport will return, although it remains to be seen if the various plans to significantly cut sonic boom noise on the ground can be realised. Programmes in the US and Japan are focused on defining acceptable sonic boom noise levels and there is growing political will to revisit the issue. As Concorde conclusively proved, there is no insurmountable challenge in building a supersonic aircraft, but to make it effective and profitable it needs to fly over land and not be restricted to only transoceanic flights.

This chapter looks at three supersonic passenger transport programmes at various stages of development, although in the case of Aerion its progress came to a shuddering halt in May 2021 when it seemed to be within reach of its supersonic dream. In the US, Spike Aerospace has relaunched its S-512 aircraft, while there are growing murmurs that Russia wants to re-enter the race with a clean-sheet Tupolev supersonic aircraft using an existing Tu-144 as a testbed.

Spike Aerospace S-512

In January 2013, US aviation and aerospace engineering firm Spike Aerospace announced it was designing a new supersonic jet, the Spike S-512, which the company claimed would cut travel times in half between major global locations. It said the first prototype would be released in 2018-2020.

By October 2013, Spike announced it had completed its initial designs for the S-512, to be configured with luxury seating for 12-18 passengers at a cruising speed of Mach 1.6. The jet would take passengers from New York City to London in three hours and from Los Angeles to Tokyo in less than seven hours, compared to seven or 11 hours on subsonic aircraft.

Regarding the issue of noise levels, as early as late 2015 Spike proclaimed that even when cruising at a supersonic Mach 1.6, the S-512

▼ Spike's S-512 will be capable of New York-London in just over three hours.
Spike Aerospace

HITS AND MISSES

would produce a nearly inaudible sonic boom of less than 70 PLdB. – a person on the ground would only hear a very soft, muffled clap when the S-512 passed overhead in supersonic flight.

Spike appeared to have solved the sonic boom issue nearly a decade ago, although with no prototype or demonstration aircraft flying what actually happens in flight has yet to be definitively proven. In 2015, Spike said it was able to design a low sonic boom aircraft by taking advantage of three factors: the aircraft's size, weight and aerodynamic configuration.

With just 18 passengers, the S-512 is significantly shorter and lighter than the Concorde, but by focusing its efforts into engineering, design and analysis Spike was able to reduce the sonic boom without affecting other flight performance characteristics. These efforts included a modified delta wing design that features a dramatically swept-inboard wing with a slender outboard section.

As with all of aviation, the COVID-19 pandemic slowed progress at Spike, but that changed in May 2025 when Spike declared it had "returned ready to reshape the future of flight." Now based in Atlanta, Georgia, Spike resumed development of the S-512 at a time when the demand for smarter, more responsive global travel was rising, heightening the case for the smaller, more environmentally responsible and relaunched S-512 Diplomat.

Spike Aerospace president and CEO Vik Kachoria said that Spike has always believed in the promise of supersonic travel and now the technology, market and regulatory climate are aligning and Spike is ready to lead: "This isn't a comeback story. It's the next step toward a future where distance is no longer a barrier to doing great things."

According to Kachoria, Spike is leveraging aerodynamic shaping to dramatically reduce the boom to a sonic thump akin to the sound of a car door closing at a distance. With a target perceived noise level below 75 decibels, the Diplomat is being designed to comply with emerging US and European noise standards.

◀ Spike Aerospace's Vik Kachoria has always believed in the promise of supersonic travel.
Spike Aerospace

Noise levels are also becoming tighter at and around airports for all aircraft. To meet these standards, Spike is using modified commercial engines and a streamlined, windowless cabin featuring panoramic digital displays fed by external cameras. The Diplomat has been designed using computational

HITS AND MISSES

▶ Interiors on the S-512 will be light-filled, spacious and luxurious. Spike Aerospace

fluid dynamics to redefine the aircraft's shape and create a sleek, optimised design that reduces drag and maximises fuel efficiency. By focusing on aerodynamic performance, Spike has increased the range and capacity of the S-512, which will give future operators greater flexibility in selecting routes than were available with Concorde services.

Today's passengers are showing a strong preference for premium cabins and it is hard to see many accepting the austere Concorde cabin when paying more than what a first class ticket on a Boeing 787 or Airbus A350 costs. To match the interior to the airplane, Spike has developed its luxury multiplex cabin, which prioritises the passenger experience with a spacious, high-tech space that replaces windows with high-definition screens spanning each side of the aircraft. These displays can show any view, whether it's the real-time aircraft surroundings, a favourite movie or a work presentation, all via a personal electronic device.

By removing the windows, the cabin offers unprecedented quietness and strength that allows passengers to relax or work in comfort and style. With bespoke leather seating, higher oxygen levels, high speed internet connectivity and lack of engine noise, passengers can immerse themselves in a world of entertainment or just set the mood with the customisable lighting and climate control systems.

When Spike relaunched the S-512 in May 2025 it entered a new phase of development with a clear vision and fresh momentum. It is actively recruiting aerodynamicists, propulsion leads and project managers, as well as growing strategic partnerships with suppliers and technology collaborators who share its vision of supersonic flight.

▼ Spike's S-512 will cruise at Mach 1.6 and connect LA-Tokyo in under seven hours. Spike Aerospace

HITS AND MISSES

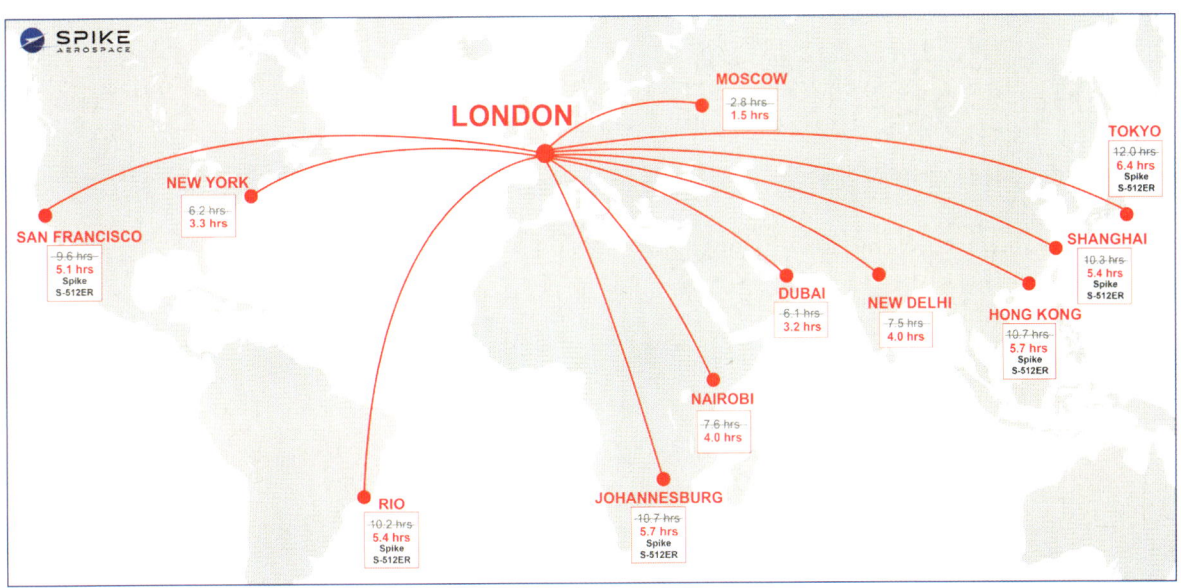

◀ **Supersonic travel transforms long-haul services into manageable trips.** Spike Aerospace

The company said its next phase will be "deliberate, but purposeful", with development milestones to be shared in time. For now, Spike is focused on building the right team, technology and momentum to ensure long-term success.

Tupolev returns

When the UK and France announced their partnership to develop a supersonic commercial transport aircraft in 1962, both the United States and the Soviet Union sat up and took notice. The threat of being left behind in the commercial supersonic race spurred both nations to announce new programmes, although it was the Soviet project that arrived first.

The Soviet Union pinned its hopes of a supersonic airliner on legendary aircraft designer Andrei Tupolev, whose Tu-144 would become the world's first supersonic transport to fly, taking to the air on December 31, 1968, two months ahead of Concorde's maiden flight. But being first was certainly not a guarantee of future Soviet success.

The Tu-144 was designed to cruise at Mach 2.0 at around 60,000ft and featured retractable canards for improved low-speed handling, a sleek delta-wing design and four powerful Kuznetsov NK-144 afterburning turbojet engines. The aircraft was intended to serve Aeroflot's high-speed long-haul routes and showcase Soviet technological progress during the Cold War, but ultimately failed on both fronts.

The aircraft initially operated mail and freight flights between Moscow and Almaty (then Alma-Aty) in Kazakhstan and did not enter commercial passenger service until November 1, 1977, on the same route. After a short but troubled and unreliable service, the Tu-144 operated its last passenger flight on June 1, 1978, less than seven months after entering passenger service.

The aircraft operated just 55 passenger services and carried just over 3,000 passengers. Despite its groundbreaking speed, the Tu-144 was plagued with persistent technical and operational shortcomings, including its limited range, high fuel consumption and deafening cabin noise. But perhaps its biggest blow was a catastrophic crash at the 1973 Paris Air Show which significantly damaged its international safety reputation.

The later crash of Tupleov Tu-144D, registered CCCP-77111, while operating a test flight on May 23, 1978, spelt the end of supersonic services. An inflight fire in the right wing forced the shutdown of two of the four engines, with a third engine subsequently failing and forcing a belly landing in a field. The aircraft had a crew of eight and all but two flight engineers survived the accident. After the official investigation, measures were implemented to prevent similar accidents and all recognised causes were addressed by adopting new methods that improved fuel pipeline durability.

HITS AND MISSES

▲ Spike's Diplomat will be configured for 12-18 passengers and is shorter and lighter than Concorde.
Spike Aerospace

However, the authorities implemented a ban on Tu-144 passenger flights, taking into account the previous Paris Air Show crash and persistent complaints about service reliability, passenger comfort and international safety concerns. This was the end of the Soviet 'Concordski' supersonic passenger transport dream, which, while shortlived, provided valuable experience and insights in supersonic passenger services.

After its removal from passenger transport operations, the Tu-144 continued flying freight services until 1983. Between 1996 and 1999 it was modified as the Tu-144LL flying laboratory in collaboration with NASA and used for high-speed flight testing from 1996. The sole remaining Tu-144 made it last flight in 1999.

In its various forms, only 16 Tu-144s were built, and with a commercial passenger transport lifespan of less than three years, it was certainly a bold, but ultimately flawed, attempt to rival the Anglo-French Concorde.

Phoenix rising

More than 40 years after the demise of the Tu-144, a new Russian supersonic passenger transport project is on the drawing board, this time at the suggestion of President Vladimir Putin.

In January 2025, the Russian newsagency TASS reported that Russia's United Aircraft Corporation (UAC) was working on a supersonic passenger jet that "may incorporate some solutions already used in the revived Tupolev-160 strategic bomber." The report said that Putin had suggested building a civilian version of a supersonic plane on the basis of Tupolev-160 after seeing the newly built strategic bomber in flight. UAC responded: "The UAC is currently working on a number of promising projects, including a supersonic passenger jet that may incorporate solutions and technologies already used in the Tupolev-160 bomber."

In November 2024, Russia's Deputy Prime Minister Vitaly Savelyev said that the development of passenger supersonic airplanes could become a reality in the near future. He noted that travel time between

HITS AND MISSES

◀ **The Tupolev Tu-144 had a short but eventful service before retiring in 1978.** Almaz Mustafin/Shutterstock

HITS AND MISSES

▲ Rumours abound that Russia is keen to launch a new Tu-160 supersonic airliner. Shutterstock/ 1826894228

▼ The distinctive shape and canards of the Tu-144. Volpati

the European part of Russia and the Far East could reach nine hours and that supersonic aircraft would significantly reduce the time, marking supersonic aircraft as the future of civil aviation.

One idea reportedly being considered is restoring a Tu-144 to airworthiness for use as a supersonic flying laboratory to support the testing of future aircraft options, particularly in terms of valuable data on safety, engine performance, efficiency and noise management for next-generation supersonic designs. Using the restored Tu-144 as a test bed, researchers would be able to directly investigate modern low-boom configurations, validate next-generation materials and engine technology, and test fly-by-wire systems in the real world, rather than relying on wind tunnel testing or simulations. The valuable data would also flow into the research of the consortium led by Russia's Central Aerohydrodynamic Institute (TsAGI) into low-boom supersonic transport technologies, currently also involving the Gromov Flight Research Institute, the Central Institute of Aviation Motors and other organisations as part of broader Russian aerospace modernisation goals.

Aerion

When British Airways operated the last Concorde service in October 2003 the future of supersonic passenger transport looked bleak, particularly given the huge cost of developing a clean sheet supersonic aircraft and the uncertain regulatory and marketing environment for supersonic travel. However, Texas billionaire Robert Bass was having none of that gloom and doom and he founded the Aerion Corporation in 2003, with the aim of commercialising supersonic aviation.

Aerion appeared to be on track in December 2020, when it started development of a new $300 million headquarters and assembly complex at Florida's Orlando Melbourne International Airport. The two million sq ft facility sited on the 110-acre Aerion Park campus was planned to house research, design, production and interior completions for the AS2

HITS AND MISSES

and future hypersonic aircraft. The assembly facility spanned the equivalent of 14 American Football fields, with flight tests in the surrounding airspace, supersonic testing corridors and offsite locations. A customer experience centre with a completions centre and a full-size cabin mock-up was planned, where customers could customise interior configuration, materials, colours and finishes.

At the time, Aerion said production of the AS2 supersonic business jet would commence in 2023, followed by a first flight in 2025 and entry into service in 2027. At the launch, chairman, president and CEO Tom Vice said it was a truly exciting day for Aerion as it launched the new home and the future of sustainable supersonic flight: "We are building the future of mobility – a future where humanity can travel between any two points on our planet in three hours or less. We will change the world and bring a new sustainable means of supersonic and hypersonic flight to reality and it will happen here at Aerion Park."

At the time Aerion also said it planned to deliver 300 aircraft over ten years of production, supported by an order backlog that now exceeded $6.5b, above the company's goal of $5b by year-end. The optimism and energy of Aerion's management at the launch of Aerion Park gave no hint of a business headed for financial oblivion, which makes what happened next all the more puzzling. Despite making significant progress on supersonic aircraft technology and securing orders for more than 90 aircraft, in May 2021 Aerion announced it was unable to secure the funding of around $4 billion it needed to certify and begin production of its AS2.

When it ceased operations, Aerion said its AS2 order book stood at $11.2b at a unit price of around $120 million per jet. The main customers included US-based fractional operators NetJets (20) and Flexjet (20), China's Avion Pacific (3) and undisclosed customer orders for approximately 50 aircraft. Following the collapse it was reported that all customer deposits would be returned.

Aerion AS2

The Aerion AS2 was never built, but the concept of a supersonic business jet still resonates within the aviation and travelling community and it has

◀ A Tupolev Tu-144 at the 1975 Paris Air Show. Gillard

▼ Aerion's AS2 made significant progress until the company abruptly collapsed in 2021. Abdulraham Own Work CC BY-SA 4

HITS AND MISSES

been selected as the model for the Spike Aerospace supersonic aircraft project. The AS2 was designed to be the world's first privately developed civil supersonic jet in more than 50 years and was aimed at the premium business aviation market.

Targeting high net worth individuals, major corporations, celebrities and heads of government and state sounded like a good strategy for a $120m supersonic business jet in a market where time is money and security and appearance are everything. However, today's world is very different to the environment when Aerion launched in 2003, which is the reason funding dried up in 2021.

The AS2 was being developed to fly at Mach 1.4, with a projected range of approximately 4,200nm, connecting New York to London in less than five hours and San Francisco to Tokyo in less than nine hours. It was designed to fly at up to 57,000ft, take off from runways of less than 7,000ft and land on runways of less than 6,000ft – a similar performance profile to today's large cabin business jets.

The AS2's structure and materials were chosen to withstand sustained cruise at altitudes above 50,000ft where skin temperatures from aerodynamic friction can reach more than 100°C. The design featured a long fuselage-to-wingspan ratio, which was critical for reducing transonic drag and increasing stability.

According to the specifications, the AS2 was to have been 148ft long, have a 79ft wingspan and be 28ft high. It was to be built from advanced carbon fibre composites and titanium, and have a supersonic-optimised delta wing with a modest sweep and laminar profile, while the tail was a cruciform with T-tail elevator for high-Mach stability. Its power was to come from three GE Affinity supersonic turbofan engines, each producing 16,000- 20,000lb of thrust. The engine had no afterburner and was built on technologies proven and refined from GE's extensive commercial engine core expertise.

When it was launched in 2018, GE said the Affinity engine employed design features and lessons learned from GE's supersonic military engines, while pulling features from its Passport business jet engine. It was optimised for prolonged high speed use using durable hot and harsh combustors and advanced coatings in the turbine section. Affinity was a twin-shaft, twin-fan turbofan controlled by a full authority digital engine control (FADEC) for enhanced reliability and onboard diagnostics. GE said it was purposefully designed to enable efficient supersonic flight over water and subsonic flight over land without requiring modifications to existing compliance regulations, adding that the engine was "designed to meet the stringent Stage 5 subsonic noise requirements and beat current emissions standards."

Mounting the engines at the rear of the AS2 reduced cabin noise and

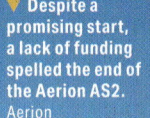

▼ Despite a promising start, a lack of funding spelled the end of the Aerion AS2.
Aerion

▲ Aerion's AS2 was designed to fly at Mach 1.4. Aerion

drag, and the central engine was embedded in the upper fuselage to maintain the clean aerodynamic shape. Affinity engines were optimised for both supersonic and subsonic cruise, which would have enabled environmentally compliant over land flight with a minimal sonic boom footprint.

Following Aerion's collapse in May, GE confirmed it had discontinued development work on the Affinity and had redeployed the team working on the engine to other programmes.

As other supersonic hopefuls emerge, it will be interesting to see if the GE Affinity can be revived on a new platform or if the lessons learned from its development can be used to design even more advanced supersonic powerplants.

SUPER POWER

Super Power

With new supersonic aircraft now on the horizon, it's time to reflect on the sheer power of Concorde's revolutionary engines and what's to come.

SUPER POWER

▲ The Olympus 593 engines were a joint Anglo/French initiative. Travelview/Shutterstock

For the first time in more than two decades, the return of supersonic passenger aircraft is becoming a reality. To make it happen, engines are being developed using advanced materials and supercomputers.

When Concorde's Rolls-Royce/Snecma Olympus 593 engines were developed, many of today's technologies were still to be invented, but that did not stop engineers building one of the finest engines ever produced.

In 2025, Boom Supersonic is shunning existing engine programmes to develop its own bespoke powerplant through collaboration with aerospace partners, while GE's experience with its Affinity engine illustrates the risks that OEMs face when backing new concepts and innovative projects.

The power behind Concorde

Concorde's arrival at any airport caused people to marvel at this wonderous aircraft that could fly faster than the speed of sound. But the real stars were not so visible, for without the four revolutionary engines Concorde could never have achieved its performance.

The engines were a living testament to the power of collaboration, which in this case transcended the sometimes prickly relationship between the United Kingdom and its near neighbour France. In the UK, Bristol Siddeley, which was purchased by Rolls-Royce in 1966, led the project with 60% of the work, while the other 40% was performed by French aerospace company Snecma. Bristol Siddeley managed all direct aspects of the engine, while Snecma looked after the rear end, including the reheat, variable nozzle, noise suppression, jet pipe and thrust reverser.

Developing the engines posed a unique set of engineering challenges. In contrast to subsonic jets, Concorde's engines had to provide efficient thrust at both take off and high-speed cruise, while also withstanding the intense temperatures and aerodynamic stresses of Mach 2 operation. To meet those challenges, the Olympus 593 engine, a direct descendant of the legendary Bristol Siddeley Olympus engine that powered the Avro Vulcan bomber, had to deliver high speed with afterburners, fuel efficiency at both subsonic and supersonic speeds and the capacity to operate reliably over transatlantic distances.

Technical stuff

The Olympus 593 was a two-spool turbojet that featured a low-pressure and -high-pressure spool and, unlike most modern jet engines which are turbofans, it was a pure turbojet. This design feature was necessitated by the need for minimal drag and to provide high efficiency at supersonic speeds.

Turbofans have large bypass ratios that make them efficient at subsonic speeds, but create excessive drag and lose efficiency above Mach 1, whereas turbojets are ideal for sustained high

SUPER POWER

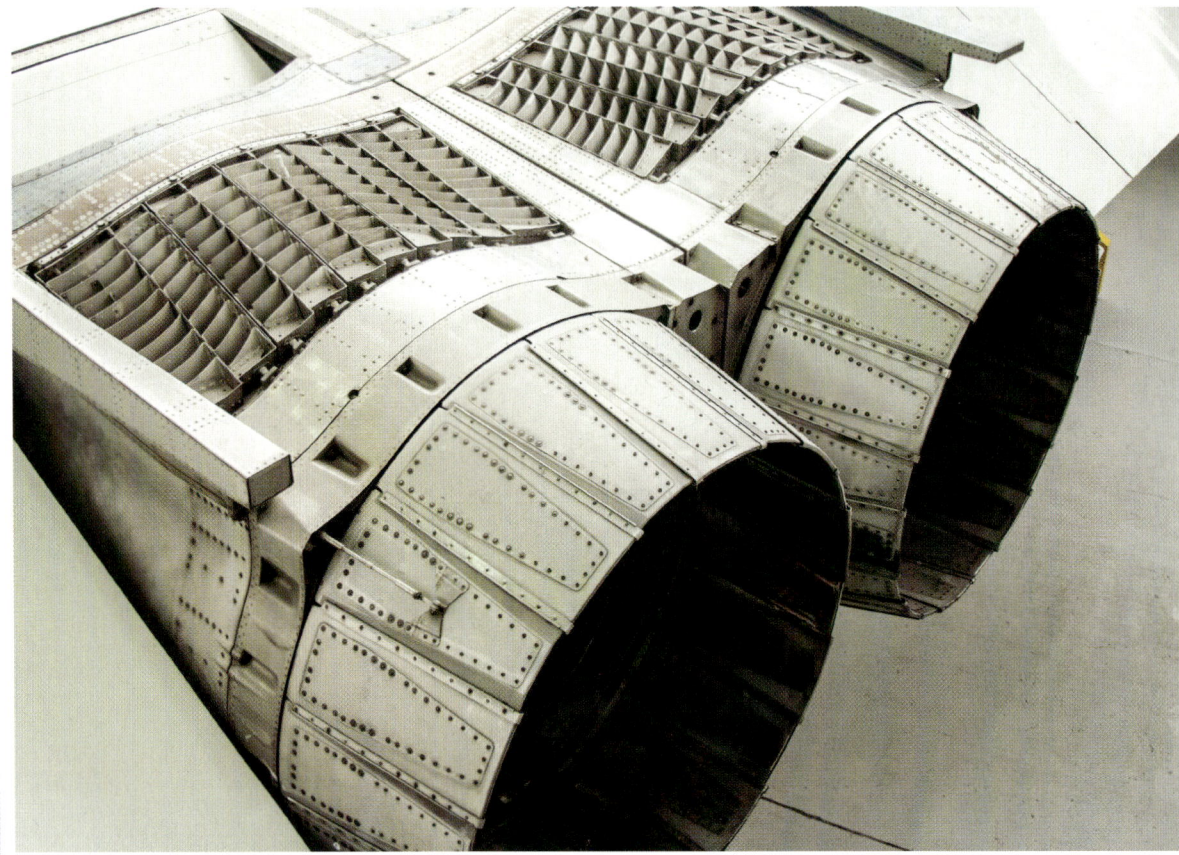

◀ Concorde's engines met the challenge of combining subsonic and supersonic flight. Piotr Przyluski/Shutterstock

▼ Boom Supersonic is opting for its own bespoke Symphony engine. Boom Supersonic

speed cruise. The 593 produced a maximum thrust of approximately 38,500lbf with afterburner (or reheat as it is known in UK terminology) and, at take-off, all four engines operate with reheat engaged, allowing Concorde to climb steeply.

The reheat function, which was critical to Concorde's performance, injects additional fuel into the exhaust stream after the turbines, which ignite it and deliver a short-term boost in thrust. It was used at take-off and the transition through the sound barrier, but was typically switched off at cruising altitude to save fuel and reduce engine wear.

While igniting exhaust gases sounds like an explosive operation, the system was engineered to come on smoothly and deliver the necessary thrust without causing excessive vibration or noise. The afterburner sections were composed of concentric rings of flame holders and spray bars designed to ensure smooth combustion and stable operation, even during the rapid airflow changes that occurred during acceleration.

Another of the many challenges was dealing with shock waves. At Mach 2, the air entering the engine inlets had to be slowed to subsonic speeds for the compressors to function effectively. The engineers solved that issue by using variable geometry intake ramps and spill doors, where the inlets adjusted their geometry to regulate airflow, minimise pressure loss and prevent

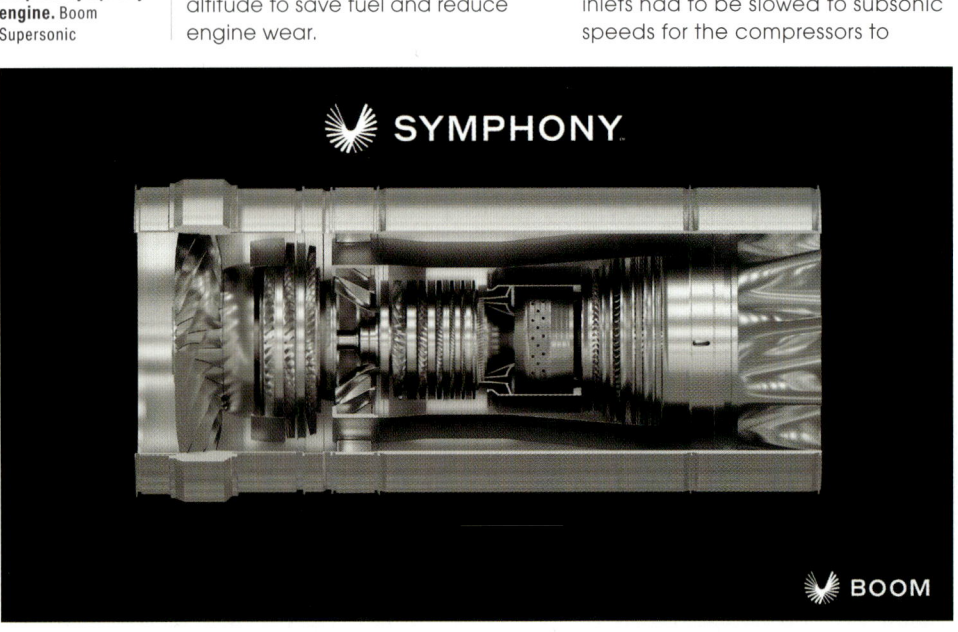

SUPER POWER

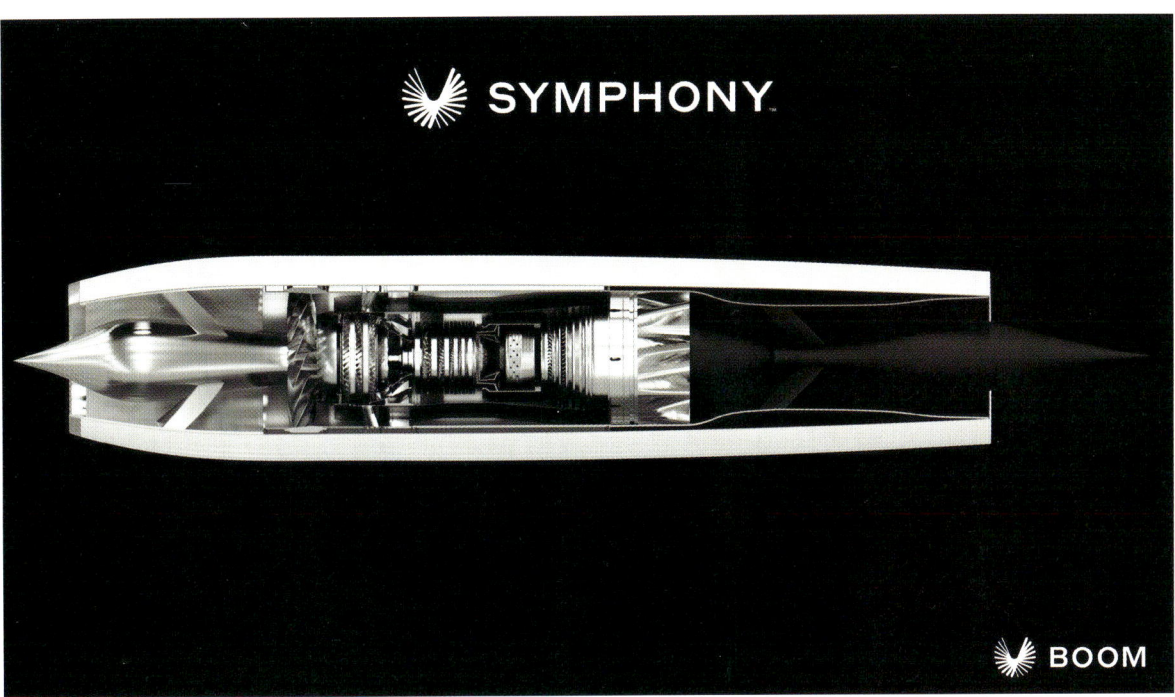

◀ The Symphony engine is a collaboration between Boom, Florida Turbine Technologies, GE Additive and StandardAero.
Boom Supersonic

compressor stalls or surges. Many would argue that the intake system was as complex as the engine itself.

Much is made of modern advanced materials. When Concorde was developed, one of the biggest engineering challenges was thermal management. At Mach 2, Concorde's outer skin could heat up to 127°C, while engine air temperatures soared even higher, so the Olympus 593 was built from high-temperature alloys that could withstand those extremes.

The engine was designed to run continuously at high power settings for long stretches, creating sustained stresses that military turbojets rarely experienced. It also had to cope with extreme altitude changes, as Concorde typically climbed from sea level to 60,000ft in less than 30 minutes, with temperatures and air pressure varying dramatically.

Fuel and the environment

To reach supersonic speeds, Concorde was an out-and-out gas guzzler, burning about 25,629 litres of fuel per hour. On the ground and at subsonic levels, the Olympus 593s consumed fuel prodigiously, but at the cruise altitude of 60,000ft at Mach 2 they became much more

▼ Boom's Overture will be powered the new Symphony engines.
Boom Supersonic

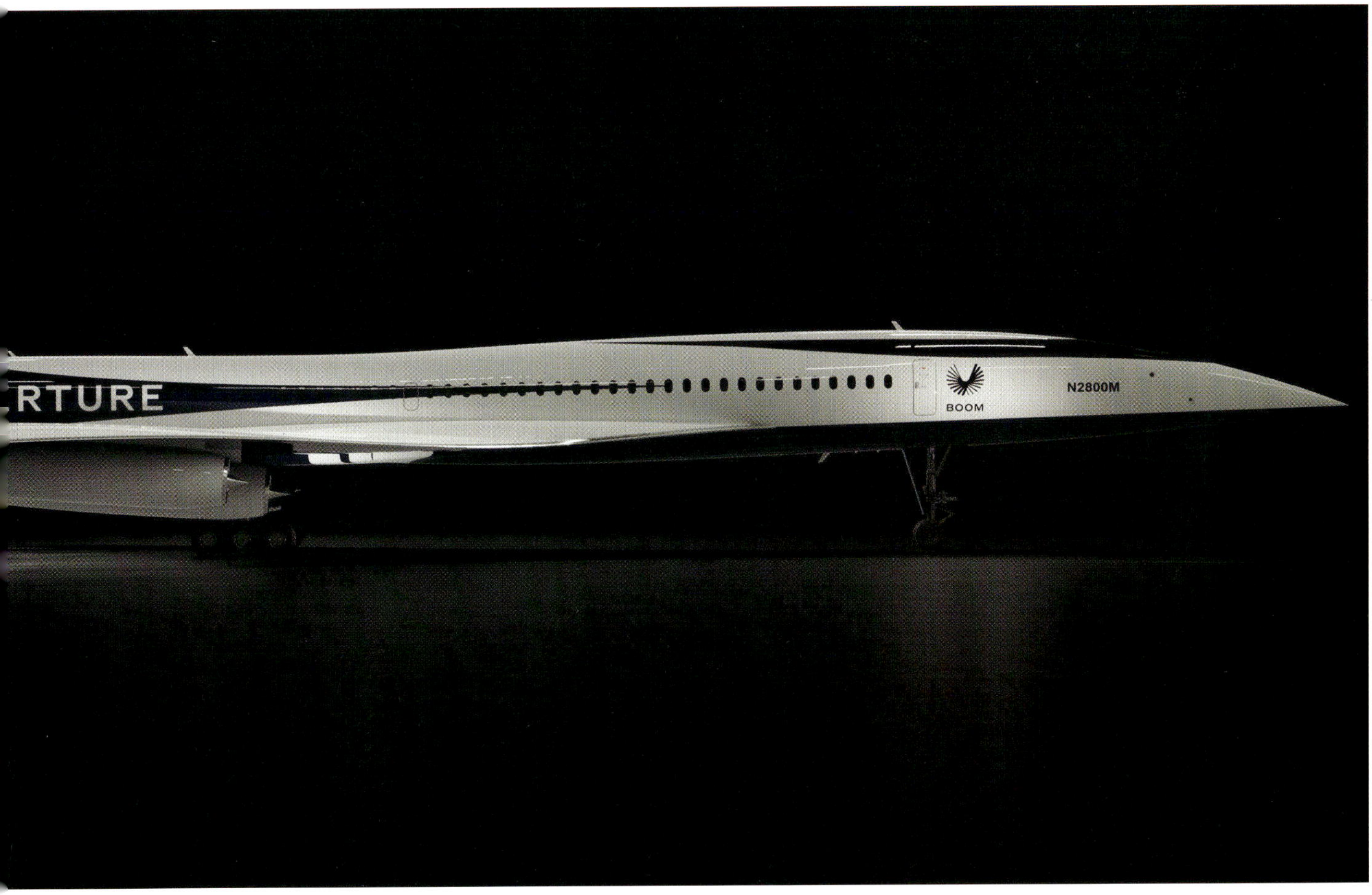

SUPER POWER

▲ Boom expects its Overture to be certified by the end of the decade.
Boom Supersonic

efficient, achieving commercially viable transatlantic crossings – at least for the luxury market that Concorde targeted.

However, the engines' thirst for fuel became one of Concorde's limiting factors, particularly as fuel prices rose steeply in response to various world shocks. Concorde also gained too much attention because of its noise, especially during take-off, when the reheat system ratcheted up the decibel readings.

Much has already been written about Concorde's sonic booms, but in an era increasingly conscious of noise and emissions, the 593 engines represented a technological compromise of pushing the boundaries of flight at the cost of significant environmental impact.

Boom Symphony

Boom Supersonic has taken a different approach to most aircraft developers by adopting a hybrid style that involves aerospace partners. The initial partners were Florida Turbine Technologies (FTT) for engine design, GE Additive for additive technology and StandardAero for maintenance.

- **Florida Turbine Technologies** is a business unit of Kratos Defense & Security Solutions and has significant supersonic engine design expertise, including key engineers responsible for the F-119 and F-135 supersonic engines that power F-22 and F-35 military aircraft.
- **GE Additive** has a proven track record of designing additively manufactured (3D printed) engine components. Using additive parts enables more streamlined development, reduces engine weight and improves fuel efficiency.
- **StandardAero** is one of the industry's largest independent maintenance repair and overhaul (MRO) providers and its input will ensure that Symphony is designed for maintainability, delivering reliable and economical operations for the life of the aircraft. Boom will also benefit from StandardAero's experience as an assembler of supersonic engines.

This unusual collaborative approach was revealed in December 2022, when Boom unveiled the Symphony propulsion system designed and optimised for the Overture supersonic airliner. Boom founder and CEO Blake Scholl said that developing a supersonic engine specifically for Overture offered by far the best value proposition: "Through the Symphony programme we can provide our customers with an economically and environmentally sustainable supersonic airplane—a combination unattainable with the current constraints of derivative engines and industry norms."

Boom described Symphony as a bespoke design leveraging proven technologies and materials to achieve optimal supersonic performance and efficiency and that t would operate at net zero carbon and meet ICAO Chapter 14 noise levels. Compared to derivative approaches, Symphony is expected to deliver a 25% increase in time on wing and significantly lower engine maintenance costs for airline customers by 10%.

SUPER POWER

▲ Concorde was built for speed, but it was a gas guzzler.
Graham Bloomfield/Shutterstock

The Symphony engine

Symphony will be a medium-bypass turbofan engine with the same basic engine architecture that currently powers all modern commercial aircraft. However, unlike subsonic turbofans, it includes a Boom-designed axisymmetric supersonic intake, a variable geometry low noise exhaust nozzle and a passively cooled high pressure turbine.

The twin-spool engine has no afterburner and is designed to deliver 35,000lb of thrust at take-off and is optimised for 100% sustainable aviation fuel (SAF). With regard to certification, Boom said that Symphony is compliant with FAA and EASA Part 33 requirements.

In November 2023, Boom announced that Symphony had successfully conducted its conceptual design review (CoDR), which is one of the key engineering milestones on the engine's pathway into service. In July 2024, the first 3D-printed parts, including fuel nozzles and turbine centre frames, had been produced and that Symphony hardware was being built and rig-tested, beginning with a full-scale test of combustor aerodynamics. Boom will conduct more than 30 engine hardware rig tests to allow for validation and optimisation of all key engine components, ranging from acoustics to combustor fuel efficiency.

A major announcement was that Boom had expanded its partnership with StandardAero to include the production of Symphony engines at the MRO's Texas facility. The Symphony assembly line will scale to produce 330 engines annually within a footprint projected to total more than 100,000sq ft of production space.

Boom expanded its engine collaboration by teaming with ATI Inc for advanced high-temperature materials and components for Symphony's high pressure compressor integrated blade and disk stages and its turbine assembly. These advanced nickel-based superalloys enable the engine to achieve high performance and reliability in sustained supersonic operation.

Wary of today's issues with new-generation engines from Pratt & Whitney, GE Aerospace, Rolls-Royce and CFM International, Boom believed it was better off starting with a clean-sheet design and an established stable of experts to bring the Symphony engine to life. Given that decision was taken three years ago, it was a bold leap of faith that appears to be paying off handsomely as the Overture heads for certification by the end of this decade.

Affinity's unfulfilled promise

Designing, developing and manufacturing a new engine is an extremely costly business, which is why engine-makers such as GE Aerospace (GE) ensure they are in solid partnerships with airframers before launching a new programme. They often strive to be the exclusive engine type available on a particular aircraft, like the Rolls-Royce Trent XWB on the Airbus A350, so they can recoup their development costs over the lifetime of a new engine programme.

SUPER POWER

▶ GE combined civil, military and business jet technology in its Affinity engine.
GE Aerospace

In 2017, GE and Aerion launched a formal process to define and evaluate a final engine configuration for Aerion's AS2 supersonic business jet. The pair had already spent two years of preliminary study on the engine, so they had a good understanding of what each could deliver. GE must have been confident in Aerion's ability to bring its supersonic jet to life.

On October 19, 2018, GE announced it had completed the initial design of the first engine.

The Affinity was a new class of medium bypass ratio engine designed to provide exceptional and balanced performance across supersonic and subsonic flights. Its DNA was a mix of the hugely popular CFM56 that powers Boeing 737s and Airbus A320ceos, GE F101/F110 military engines and the Passport business jet powerplant to create a twin-shaft, twin-fan turbofan controlled by a next-generation full authority digital engine control (FADEC) for enhanced despatch reliability and onboard diagnostics. GE said that it was "purposefully designed to enable efficient supersonic flight over water and efficient subsonic flight over land without requiring modifications to existing compliance regulations."

The Affinity engine featured:

- Efficient performance throughout the full flight envelope with a high-altitude service ceiling of 60,000ft
- An advanced twin-fan with the highest bypass ratio of any supersonic engine
- A special, non-augmented supersonic exhaust system
- A proven engine core adapted from GE's commercial airline portfolio with billions of successful and reliable hours of operational service
- A durable combustor with advanced coatings for sustained high-speed operation
- Advanced acoustic technology designed to meet or exceed regulatory requirements
- GE's additive design & manufacturing technologies to optimize weight & performance

While Affinity might have been the first civil supersonic engine in

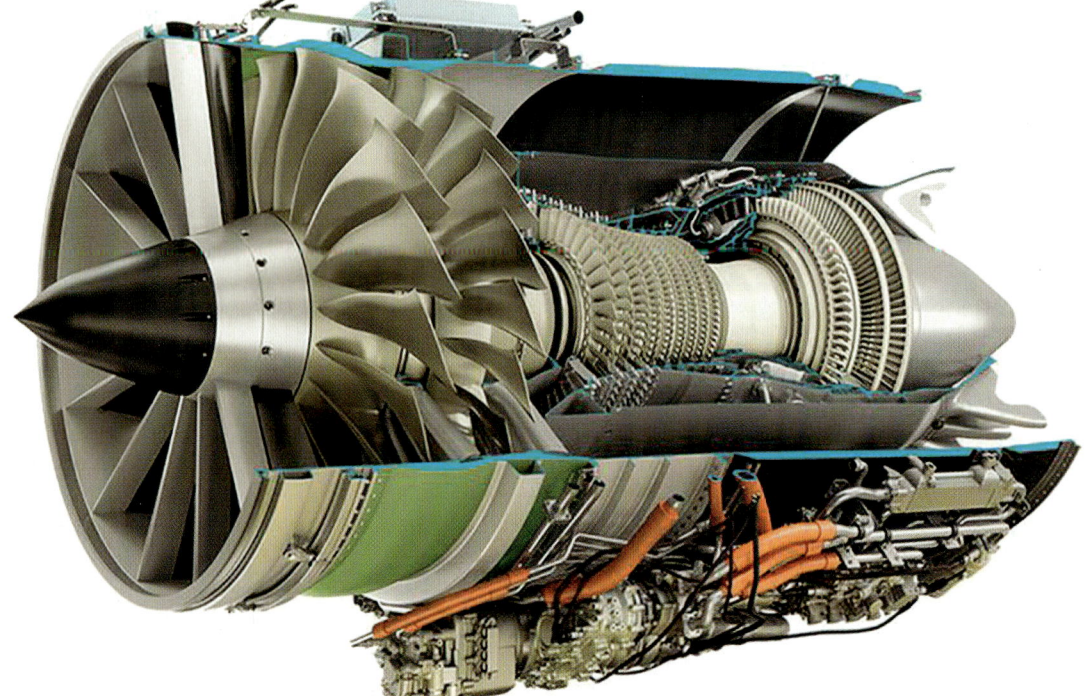

▼ GE's Affinity was developed to power the Aerion supersonic bizjet.
GE Aerospace

SUPER POWER

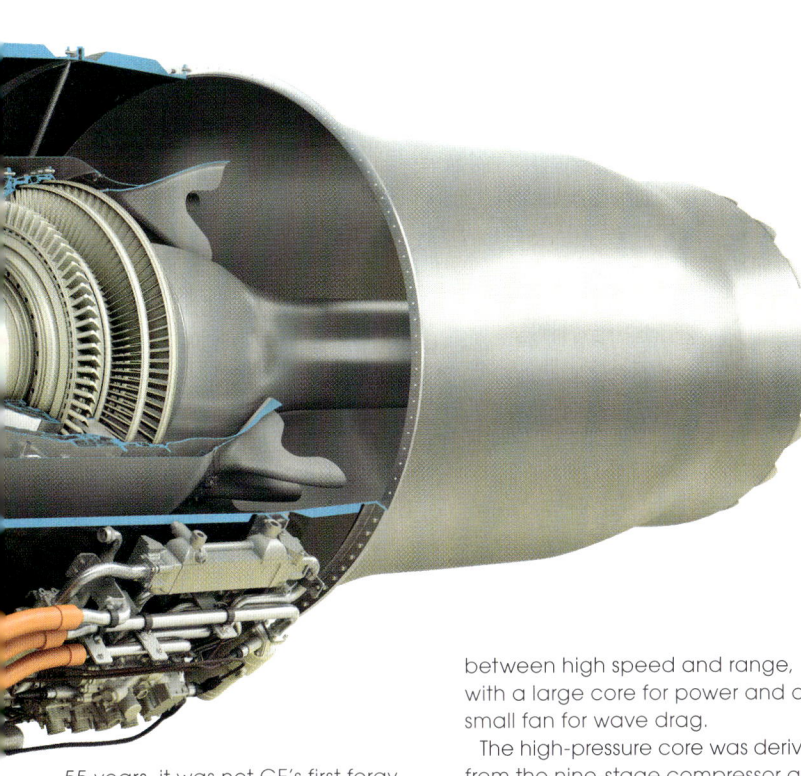

55 years, it was not GE's first foray into this field. More than 70 years previously, in the mid-1950s, it had developed the J79 engine for the Lockheed F-104 Starfighter military aircraft.

The challenge for GE with Afinity was that it needed a configuration accommodating a reasonable requirement for supersonic speed, subsonic speed and noise levels, while managing the high intake temperatures at high altitudes. In essence, Affinity was a compromise between high speed and range, with a large core for power and a small fan for wave drag.

The high-pressure core was derived from the nine-stage compressor and single-stage turbine of the CFM56, matched to a new low-pressure section optimised for supersonic speed with an 133cm diameter fan instead of the 155-173 cm fan used in the 6:1 bypass ratio CFM56. The engine's combustor incorporated advanced coating and additive manufacturing techniques and did not have an afterburner, which older supersonic designs used for cruise or acceleration, instead relying entirely on dry thrust.

High-temperature alloys and advanced coatings allowed for sustained operation at supersonic cruise temperatures without excessive cooling requirements, while improved sealing, lubrication and thermal barrier coatings increased durability at high speeds. In a forward-looking move, GE designed Affinity to be fully compatible with 100% SAF blends in anticipation of environmental demands.

Aerion and its AS2 programme came to an abrupt halt in 2021, when it was unable to raise the finance needed to produce, certify and deliver the new supersonic aircraft. GE was quite circumspect and only said that it had discontinued work on the Affinity engine and that engineers on the project had been reassigned with no impact on employment numbers.

The collapse of Aerion and the cancellation of the Affinity programme highlighted a challenge in reviving supersonic passenger transport in that even technically feasible propulsion systems require a committed airframe partner with a stable business case to reach completion. Despite its considerable financial outlay when Aerion folded, GE had no other viable market for the Affinity engine, so making further investment in the programme couldn't be justified, particularly given the high certification costs and the limited number of aircraft that would be produced.

▼ Aerion's collapse left GE's Affinity with nowhere to go.
GE Aerospace

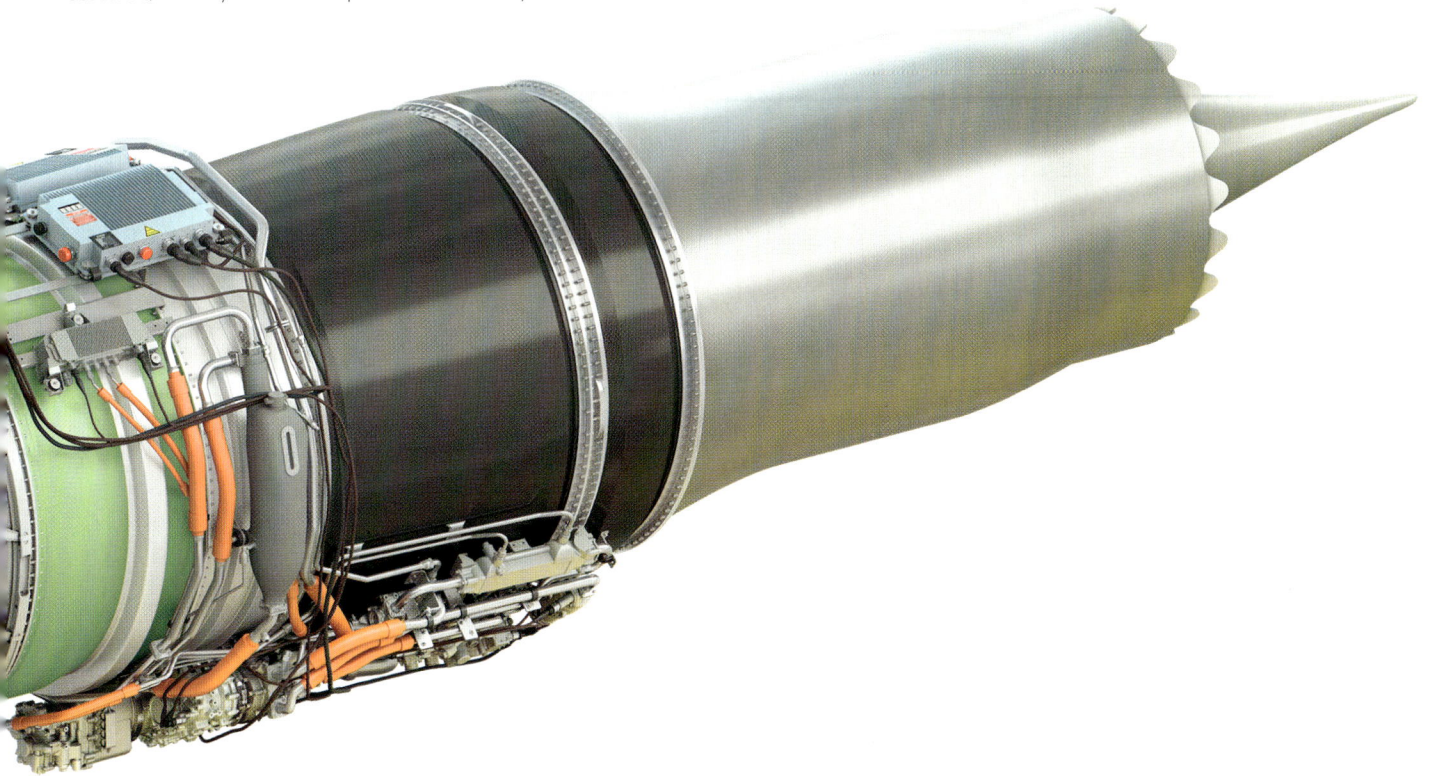

The World's Fastest Growing Aviation Website

Join us online

SUBSCRIBE TODAY!

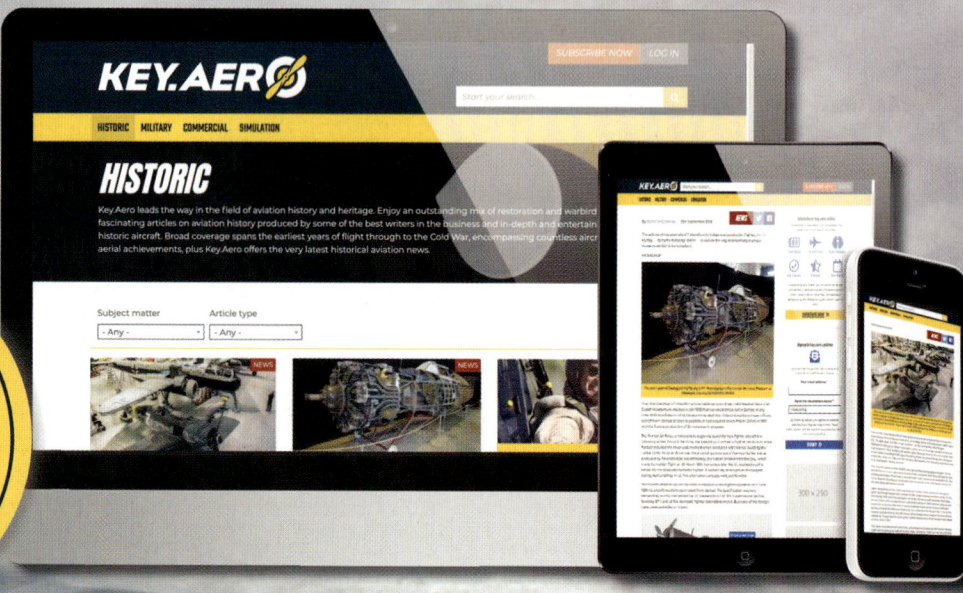

"In-depth content and high-quality photography" — **Stew**

"Well worth the money" — **Andy**

"A great place for aviation geeks!" — **Kenneth**

GREAT REASONS TO SUBSCRIBE

- In-depth articles, videos, quizzes and more, with new material added daily
- From historic and military aviation to commercial and simulation – *Key.Aero* has it all
- Exclusive interactive content you won't find anywhere else
- A fully searchable archive
- Access to all the leading aviation magazines
- Membership to an engaged, global aviation community
- Access on any device – anywhere, anytime

www.key.aero

Subscribe FROM JUST £5.99 for unlimited access

THE DESTINATION FOR AVIATION ENTHUSIASTS

Visit us today and discover all our latest releases

LITTLE GRAND TOURER
Pipistrel Explorer: comfort, 153mph cruise speed and airliner-style avionics

NORTH CAPE ADVENTURE
North through Denmark, Sweden, Finland and Norway (in a 65hp Jodel!)

SOLO IN A WWI REPLICA
Stretching your comfort zone flying a single seater

JUST £5.99 When you order direct

Order today from our online shop...
shop.keypublishing.com/collections/pilot-magazine

Call +44 (0)1780 480404 *(Monday to Friday 9am - 5.30pm GMT)*

*Free 2nd class P&P on BFPO orders. Overseas charges apply.

REGULATING THE BOOM

Regulating the Boom

The ban on over land supersonic flight is in the spotlight with the US President ordering it should be changed. But how are these rules formulated?

REGULATING THE BOOM

As aircraft and propulsion technologies rapidly advanced in the mid-20th Century and more long-haul routes became viable, airline executives, aerospace engineers and the travel industry looked to supersonic flight to reduce the tyranny of distance. With the promise of drastically reduced flying times – such as an eight-hour flight between London and Sydney, Australia – the interest turned into action with the development of the Anglo-French Concorde and the US Boeing 2707 prototype. While the US project ultimately proved unviable, Concorde entered service in 1976, although its success was already jeopardised by the ban on supersonic flight over land.

This restriction arose primarily from the unacceptable community impact of sonic booms, which proved far more disruptive than early proponents of supersonic passenger transport had anticipated. A sonic boom is not simply louder jet noise, it's a sharp N-wave pressure signature generated when an aircraft exceeds Mach 1. Unlike subsonic noise that diminishes predictably with distance, the boom propagates to the ground as a focused shockwave covering large areas. Even at cruising altitudes of 50,000–60,000ft, over-pressures of 1-2psf can damage structures, shatter windows and startle populations.

Perhaps the most decisive blow for supersonic flight followed the FAA/USAF Oklahoma City sonic boom experiment in 1964, which subjected the city to eight daily booms over six months. The programme was designed to study public acceptance and gather acoustic data, but the tests triggered intense backlash, with around 15,000 damage claims lodged by residents whose properties suffered broken windows, cracked plaster and other structural impacts, while other people reported persistent psychological stress from the sonic booms. These claims triggered hundreds of lawsuits and lasting political fallout which ultimately led the US authorities to take action in 1973.

ICAO's noise limitations

The International Civil Aviation Organization (ICAO) is a United

▼ Concorde never lived up to its full potential once over land flights were banned. NARA

REGULATING THE BOOM

▶ **The United States proposed going supersonic with the Boeing 2707.** Donald Huebler

Nations agency that helps 193 countries to co-operate and share their skies for their mutual benefit. It was established in 1944 and has helped nations to establish a dependable network of global air mobility, connecting families, cultures and businesses, as well as promoting sustainable growth and socio-economic prosperity wherever aircraft fly. ICAO's five strategic objectives comprise safety, air navigation capacity and efficiency, security and facilitation, economic development of air transport and environmental protection.

The last of these objectives focuses on minimising the adverse social effects of civil aviation and fosters ICAO's leadership in all aviation-related environmental activities. It is consistent with UN environmental protection policies and practices.

▶ **ICAO is developing new supersonic noise regulations.** KI Photography/ Shutterstock

▼ **Moves are afoot in Washington to scrap bans on over land supersonic flights.** Bahammou1/ Shutterstock

Annex 16 to the Chicago Convention is the international standard for environmental protection in aviation. It evolved through amendments agreed by ICAO's Committee on Aviation Environmental Protection (CAEP) and is split into Volume I (Aircraft Noise) and Volume II (Aircraft Engine Emissions), with a more recent Volume III (CO2 Certification) and Volume IV (CORISA – Carbon Offsetting and Reduction Scheme for International Aviation).

ICAO Annex 16, Volume I: Aircraft Noise (2025) sets certification noise limits for civil aircraft types and compliance is mandatory for all states party to the Chicago Convention, which is the treaty signed in Chicago on December 7, 1944, that established the International Civil Aviation Organization. Aircraft are certified to specific noise standards called 'Chapters', each defining the maximum permitted noise levels measured at three standard certification points: flyover, sideline and approach. Of most relevance in 2025 are Chapters 2, 3, 4 and 14, with the last being the latest and most stringent for subsonic jets.

REGULATING THE BOOM

Chapter 3 is an older baseline standard for 1977/1978 onwards, with many existing aircraft certified to this , whereas newer types must meet the stricter Chapter 4 rules adopted in 2001. These require a cumulative margin of 10EPNdB (Effective Perceived Noise in Decibels) below Chapter 3 limits, with most modern subsonic jetliners meeting or exceeding Chapter 4 standards.

Chapter 14 was adopted in 2013 and applies to new large subsonic designs from 2017, requiring a cumulative margin of 17EPNdB below Chapter 3 limits. It called for a phased application to existing types under production from 2020. In 2025 new subsonic transport jets must comply with Chapter 14.

Subsonic jets over 8,618kg MTOM and propeller-driven aircraft over 8,618kg must meet relevant Chapter standards, while helicopters have separate limits set out in Annex 16 Volume I (Chapter 8 and 11). Annex 16 still includes no new noise standard for supersonic aircraft but, as of 2025, the ICAO, FAA and EASA are working on future standards.

For now, there is no Chapter equivalent for new supersonic airliners. Over land sonic booms remains prohibited in many countries, pending development of low-boom standards. ICAO is researching sonic boom acceptability and low-boom acceptability metrics, but no finalised regulations have yet been promulgated.

Measuring noise

Boom Supersonic is well advanced with developing the airframe and propulsion technology it will use for supersonic flight and is also working hard on overcoming the noise limitations that bedevilled Concorde's efforts to gain widespread adoption. As the sonic boom noise is such an important element, it is worth understanding how ICAO defines and measures it as this will have a significant effect on future changes in regulations.

The ICAO's standard methodology is built around the EPNdB metric that is in use for subsonic aircraft, which now sits at a cumulative margin of 17EPNdB below Chapter 3 limits. To satisfy lawmakers in the US, it is likely that the same certification level will be placed on supersonic flight over land, along with other noise limits for take-off and landing.

EPNdB accounts for frequency weighting, which reflects human perception of pitch, tone corrections for engines and the duration of the noise event. It is specifically designed to correlate well with perceived noisiness by communities near airports.

Chapter 14, like Chapters 3 and 4 before it, requires certification tests at three standard locations: flyover, sideline and approach. The flyover phase measures take-off climb noise when an aircraft at full take-off thrust passes over a microphone 6.5km from brake release.

The sideline phase uses microphones placed 450m from the runway centreline to measure lateral noise during take-off. Approach noise is measured from an aircraft configured for landing on a 3° descent path using a microphone located 2km from the runway threshold.

Chapter 14 has moved from standards with fixed point limits to defining limits relative to Chapter 3 levels. The key rule is that an aircraft's cumulative noise level, which is the sum of the flyover, sideline and approach levels, must be at least 17EPNdB below the cumulative Chapter 3 limit. For example, a large subsonic jet with a MTOW greater than 400,000lb might have a Chapter 3 cumulative limit around 315EPNdB, whereas under Chapter 14 the same aircraft's

◀ While Concorde ushered in a new era, its sonic boom held it back. NARA

▼ When Concorde entered service, noise abatement rules were less stringent than today. NARA

REGULATING THE BOOM

▶ While Concorde had the glamour, BA's jumbo jets could operate globally.
British Airways

▼ New supersonic jets will also have to adhere to subsonic noise limits around airports.
British Airways

REGULATING THE BOOM

cumulative measured noise must be less than 298EPNdB.

However, Chapter 14 does not impose a fixed reduction at each individual measurement point. This allows manufacturers to trade margins between flyover, sideline and approach measurements, so long as the cumulative margin is reduced by greater than 17EPNdB., leaving OEMs free to find reductions where they can, such as switching to quieter new-generation engines that reduce flyover noise or redesigning flaps or landing gear doors to reduce approach noise.

No speed limits

The existing US ban on supersonic flights over land is of real concern not just to Boom Supersonic, but also to other programmes under development in the US and globally. Given the 2025 push by President Donald Trump's administration to promote technology development in the US, it is unsurprising that politicians are eager to have the speed ban overturned.

In May 2025, US Senator Ted Budd and aviation subcommittee chair Troy Nehls introduced the Supersonic Aviation Modernization Act (SAM), which calls on the FAA to update or issue new regulations

▲ It is ironic that Concorde suffered most opposition in New York, where it is now on public display. Dmitry Morgan/Shutterstock

▼ Boom's Blake Scholl said a new age of faster travel has been stifled. Boom Supersonic

REGULATING THE BOOM

▲ **Boom Supersonic is pressing ahead with a first flight by the end of the decade.** Boom Supersonic

within one year, allowing civil aircraft to exceed Mach 1 speeds, with the proviso that no sonic boom reaches the ground within the US. The lawmakers say that the 1973 restriction is outdated and stifles innovation in the aerospace industry and that the SAM Act would open the door to next-generation aviation by enabling quiet supersonic flight while maintaining safety and noise standards. When announcing the proposed change, Nehls said: "Our nation's laws and regulations should encourage these innovations and uplift companies that are leading in industries, including the aviation industry. My legislation cuts regulatory red tape without minimising safety, and incentivises further innovation in the aviation industry, helping America remain competitive and the envy of the world."

While nothing is certain in politics, the bill has received strong backing from industry leaders, including Boom Supersonic and the National Business Aviation Association. The impetus for change gained momentum in January 2025, when Boom's XB-1 demonstrator became the first privately developed civil aircraft to break the sound barrier over the continental US.

When the ban was originally enacted on April 27, 1973, the FAA effectively placed a speed limit in the sky, which, in April 2025, Boom Supersonic CEO Blake Scholl said crippled progress in flight and unintentionally stifled American innovation in aviation. Scholl said: "We also lost something deeper – the economic effect and cultural benefits of faster travel. Getting to your destination faster is not just a matter of time savings, it can be the difference between going and not going. In the absence of a new age of faster travel, innovation has been stifled, leaving us in what I believe is the dark age of commercial flight, (where) we have gone more than six decades without a mainstream speed-up in flight."

Unsurprisingly, Scholl had plenty to say about the need for the SAM bill to be introduced and for lawmakers and the industry to focus on a noise-based approach rather than limiting speed. He believes that shift would align with the original intent in 1973, which was protecting the public from "bad sounds", while enabling a framework for innovation: "Clearly we should

REGULATING THE BOOM

allow supersonic flight that does not create audible sonic booms. Then we can go further—a noise standard would allow even faster flights without objectionable noise. This would allow manufacturers to develop and test new supersonic aircraft, fostering a competitive market at a time when maintaining US leadership in next-generation aerospace is critical. As the country looks to reindustrialise, we cannot hold back our own progress with a regulation that never should have existed. It's time to say goodbye to the supersonic ban and embrace a better future of speed and innovation."

Trump goes supersonic

On June 6, 2025, President Donald Trump signed an Executive Order (EO) to promote supersonic aviation in the United States and "remove regulatory barriers so that US companies can dominate supersonic flight once again."

The EO directs the administrator of the FAA to repeal the prohibition on over land supersonic flight, establish an interim noise-based certification standard and repeal other regulations that hinder supersonic flight. The FAA would need to repeal the prohibition on over land supersonic flight in 14 CFR 91.817 within 180 days and issue a Notice of Proposed Rule Making to establish a standard for supersonic aircraft certification within 18 months of the order.

It also instructs the administrator to establish a standard for supersonic aircraft noise certification that considers community acceptability, economic reasonableness and technological feasibility. The EO added that advances in aerospace engineering, materials science and noise reduction now make supersonic flight not just possible but safe, sustainable and commercially viable. The document also points out that American companies developing supersonic aircraft have already entered into government contracts and agreements with major commercial airlines, such as United Airlines and American Airlines, who have committed to purchase supersonic jets to enhance their fleets with faster travel options.

The timing and wording of the Executive Order largely mirror an opinion piece released by Scholl on April 27, 2025, and parts of the SAM Act released in May 2025. These three documents were issued within weeks of each other and demonstrate that the push to repeal the supersonic speed rule is a well co-ordinated, sustained effort that will likely achieve the outcome that all three are calling for.

Concorde's workarounds

While most of the noise issues around supersonic flight focus on the over land sonic boom, there are still the issues of subsonic flight noise at and around airports and populated areas that any new entrant must overcome. When Concorde entered service, noise abatement issues were not as strict as they are today, but even then Concorde could not meet the ICAO requirements. On take-off with afterburners engaged, Concorde's noise levels significantly exceeded the ICAO Chapter 2 limits that applied to new subsonic jets in the 1970s.

After heavy political pressure was brought to bear and legal arguments preventing Concorde operating were defeated, stringent noise abatement procedures were placed on the airliner at take-off and landing. It also faced strict noise monitoring, particularly at Heathrow and JFK, but the biggest hurdle came from the Port Authority of New York and New Jersey, (PANYNJ), the airport's operators.

In the 1970s, aircraft noise issues over populated areas became a serious political issue and, in 1976, PANYNJ attempted to ban Concorde outright from JFK, citing local environmental and community objections. British Airways and Air France, supported by the UK and French governments, sued to overturn the ban, leading to a high-profile legal battle in the US federal court. The airlines argued the ban was discriminatory and violated US obligations under international aviation agreements. In 1977, a federal judge ruled the PANYNJ ban was "arbitrary and capricious", ordering JFK to accept Concorde flights. That decision established that a local airport authority couldn't unilaterally exclude a federally certified international aircraft if it met national safety rules, even if it was noisier than typical subsonic airliners.

Despite the legal victory, airports around the world were facing

▼ Air France received strong national support for Concorde flights. Travelview/Shutterstock

REGULATING THE BOOM

▲ **Concorde offered speed on a few routes, but the 747 changed the face of mass travel.** Eliyahu Yousef Parypa/Shutterstock

▶ **Boom's Overture is developing technology that stops sonic booms from reaching the ground.** Boom Supersonic

increased community pressure to reduce noise impacts from all jet airliners, with Concorde singled out as the worst of the worst. In Queens and Brooklyn near JFK, residents organised vocal opposition, arguing Concorde's noise would worsen the quality of life and property values, while Heathrow faced similar opposition in West London. Washington/Dulles and Paris/Charles de Gaulle also faced the backlash, but their locations in more lightly populated areas meant the scrutiny was far less than at JFK and Heathrow. However, all four airports did impose strict noise abatement procedures for Concorde operations, with a focus on take-off procedures, approach and landing, scheduling restrictions and airport slot limits.

While the exact conditions varied, JFK, Heathrow and Dulles airports all consistently employed a set of baseline noise abatement techniques for Concorde flights. At take-off, afterburners could only be used until safe climb was established, typically within 60-90 seconds of lift-off and using very steep initial climb gradients to gain altitude quickly over populated areas. Concorde employed continuous descent profiles on approach to minimise level segments requiring engine spool-up and maintained idle or

REGULATING THE BOOM

near-idle thrust for long stretches to reduce landing noise. Heathrow also established noise preferential routes for Concorde, requiring it to fly designated corridors to reduce the impact on populated areas.

The final restriction was the frequency and timing of flights, with operations limited to specific daytime hours to avoid disturbing communities at night, which became essentially mandatory at all airports that handled Concorde flights.

Public reactions

While each of the main four airports handling Concorde had their own nuanced set of regulations, the reality was that all demanded it use afterburner cutbacks and steep climbs to minimise noise, with JFK imposing the strictest rules and Heathrow also having strict limits. Washington/Dulles and Paris/Charles de Gaulle airports were not as heavily restricted.

At JFK, Concorde was limited to two daily flights by Air France and British Airways, while at Heathrow Concorde flights were restricted to daytime slots. The upside was that although limiting flights reduced flexibility and expansion opportunities, scheduling Concorde during premium daylight hours also ensured that British Airways maximised its yields from business travellers. Washington/Dulles, the only other US airport to see regular scheduled Concorde service, was built with more buffer land and, with services only operating a few times a week, opposition from the public and politicians was more muted. Air France's home base at Charles de Gaulle enjoyed more public support to support French aviation prestige, but local authorities required noise-abatement procedures on departure and approach. These measures added operational complexity and reinforced Concorde's image as an elite, limited-frequency service.

The result of these operational limitations, including the rules barring supersonic flights over land, ensured that Concorde remained a small-volume, premium priced service, although it was highly valued by its frequent business travellers. However, the inability to add more flights or open up new destinations without lengthy noise abatement processes meant the economics of Concorde was always under the microscope.

Prospective new entrants, such as Boom, are working on technologies to avoid sonic booms reaching the ground, but they will also have to meet the latest and far stricter ICAO take-off and landing standards than those applied to Concorde. With commercial aviation increasingly in the sights of environmentalists and lawmakers, there will also be serious questions around sustainability and carbon emissions that will add another layer of regulatory complexity.

▲ Long-haul airlines like Qantas were never able to fly Concorde due to over land bans.
Qantas

HYPERSONIC PROMISE

Hypersonic Promise

Hypersonic passenger flights are decades away, but important airframe and engine research is validating technologies to make it a reality.

When Chuck Jaeger first flew faster than the speed of sound in 1947, he opened the door for designers and engineers worldwide to search even more earnestly to unlock hypersonic flight. Those innovators encountered a unique set of challenges to overcome as they encountered new aerodynamic phenomena at hypersonic speeds, such as shockwaves, extreme heating and high dynamic pressures that needed entirely new approaches to aircraft design and materials science.

When an aircraft or other space vehicle exceeds Mach 5, it is in the realms of hypersonic flight. This flight domain is characterised not only by velocity but also by unique physical phenomena that emerge due to the extreme kinetic energy and aerodynamic heating involved. It differs from subsonic or supersonic because hypersonic flows exhibit significant dissociation and ionisation of atmospheric gases that lead to complex interactions between the vehicle's surface, boundary layer and shockwaves.

Such extreme speeds mean that thermal management is a primary challenge for engineers. As vehicles compress air ahead of them, temperatures can soar to more than 2,000°C, well beyond the tolerance of conventional materials. The only way to protect the surfaces from structural degradation is to use advanced ceramics, ablative coatings and thermal protection systems.

Aerodynamics

To deal with the intense shockwave formation, hypersonic and

▼ **Hermeus aims to produce a new Quaterhorse vehicle every year.** Hermeus

HYPERSONIC PROMISE

high-drag hypersonic vehicles are usually designed with slender, wedge shaped profiles to manage the heat flux and minimise wave drag. Conventional turbojet engines are ineffective at such speeds, so hypersonic vehicles generally use rocket propulsion, supersonic combustion ramjets (scramjets) or hybrid systems. Hypersonic propulsion relies on air-breathing machines that effectively combust fuel in high-velocity airflow without carrying oxidiser. These engines only function within a specific speed envelope, requiring innovative multi-mode propulsion strategies for acceleration to and deceleration from hypersonic velocities.

The hypersonic field can be a maze of daunting technical demands that brings together a cutting-edge blend of materials science, propulsion engineering, thermodynamics and aerodynamics. Hypersonic flight applications span both military and civilian sectors, with current programmes focusing on high speed weapons, reconnaissance platforms and experimental aircraft, while future concepts also envision ultra-fast passenger travel.

Hypersonic transcontinental travel would reduce flight times from hours to minutes, which has generated interest from researchers and organisations aiming to build the next generation of passenger transport aircraft. It will be a difficult nut to crack, with significant technical and economic hurdles that are likely to take decades of further research and investment to overcome.

Hermeus horses

Hermeus is a US aerospace and defence technology company founded to radically accelerate air travel by delivering hypersonic aircraft. The company aims to develop these quickly and cost-effectively by integrating hardware-rich, iterative development with modern computing and autonomy.

Hermeus is exploring several aircraft concepts, with its first major demonstrator being the Quarterhorse, a remotely piloted vehicle designed to validate its signature Chimera engine at hypersonic speeds. The ultimate goal is to create passenger aircraft capable of reducing transcontinental and transoceanic flight times to just a few hours, making routine hypersonic travel a reality.

Quarterhorse MK 1

In March 2024, Hermeus unveiled its first first aircraft that later debuted in May 2025. The Quarterhorse Mk 1 was designed, built and flown in just seven months and was the company's second fully integrated vehicle following on from Quarterhorse Mk 0, which completed its test campaign in November 2023.

Quaterhorse Mk 1 (Q1) is an uncrewed, remotely piloted aircraft powered by a GE J85 engine and its primary mission is to demonstrate high-speed take-off and landing, which is a key enabling capability unique to future Hermeus hypersonic aircraft.

In the official announcement, Hermeus vice president of test, Don Kaderbek, said that moving into the integrated test programme was the culmination of a huge team effort and a significant emotional event for the entire company: "As we begin the journey to first flight, we will conduct a comprehensive evaluation of the aircraft's performance while simultaneously examining the effectiveness of our test procedures, safety culture and interdisciplinary team collaboration. We're excited and humbled to conduct this testing at the legendary Edwards Air Force Base."

Each aircraft in the Quarterhorse programme progressively increases in complexity

HYPERSONIC PROMISE

▲ Hypersonic flights will link London-New York in around one hour.
Andrey l/Shutterstock

▶ Hermeus is building it Quarterhorse Mk 2 at its Atlanta HQ.
Hermeus

and builds on the learnings from its predecessor. Hermeus believes this approach manages risks across multiple vehicles and accelerates delivery of products and services to its customers. So far, it has proven successful in delivering significant improvements in the capabilities of rockets, satellites and drones, and now Hermeus is bringing that power of iteration speed to its aircraft.

At the unveiling, Hermeus also announced that its next iteration, Quarterhorse Mk 2, will feature the Pratt & Whitney F100 engine and fly at supersonic speeds in 2025. For more than 50 years, the F100 engine has powered the F-15 and F-16 fighter jets with industry-leading reliability and now it is entering this new era of aviation.

First flight

The pace picked up on May 27, 2025, when Hermeus announced the first flight of its Quarterhorse Mk 1, taking the aircraft from clean sheet to flight ready in a little over one year. This was a significant milestone in the company's development of a high-Mach and hypersonic aircraft, while also advancing Hermeus' mission to operationalise hypersonic technologies. The announcement revealed no specifics about the flight profile or aircraft performance, but it said data from the campaign had validated design and performance models, including aerodynamics, stability and control. The primary focus was to validate high-speed take-off and landing because the aircraft's unique configuration, driven by high speed flight, makes these basic operations distinctly challenging.

Testing also validated the performance of vehicle subsystems, including propulsion, fuel systems, hydraulics, power, thermal management, avionics, flight software, telemetry, flight termination and command and

▶ Hermeus is developing a new concept engine named Chimera.
Hermeus

HYPERSONIC PROMISE

◀ Hermeus Quarterhorse Mk 1 during high-speed taxi trails. Hermeus

control. Hermeus co-founder and president Skyler Shuford explained: "The real-world flight data from Mk 1 provides significant technical value that we're rolling into our next aircraft. Moreover, the team has accomplished this milestone on a challenging timeline while operating within the overall aerospace system – all to support rebuilding America's lost capability to quickly develop brand-new, full-scale jets."

The Hermeus approach emphasises hardware richness, which it defines as building multiple prototypes in quick succession that allow the team to take well managed technical risks. The company said that while driving technical progress towards high-speed flight this approach simultaneously enables development of its team and talent.

Aiding that development is the one-aircraft-a-year cadence that drives Hermeus engineers and technicians through multiple crucibles of full life-cycle aircraft development in a very short period of time. The outcome is that the company is progressively building a team capable of solving the hardest engineering challenges of aviation to operationalise hypersonic aircraft.

Not resting on its laurels, Hermeus is now actively reviewing data and integrating the lessons learned into the next iteration, Quarterhorse Mk 2, which is being manufactured at its Atlanta headquarters. The Mk 2 is a high-Mach aircraft the scale of an F-16 and is designed to de-risk uncrewed supersonic flight. It is on track to fly later in 2025.

Chimera

Chimera is the world's first commercially developed turbine-based combined cycle engine. At low speeds, Chimera is in turbine mode, just like any jet

▼ A possible timeline for aviation's hypersonic future.

HYPERSONIC FLIGHT DEVELOPMENT TIMELINE

Foundations 1ornets	Early Air-Breathing Experiments	Tecchnology Consolidation	Hyprersenic weapons and passenger concepts	Looking Ahead
Ramjet and scramjet research	Ground-tested scramjet engines	X-43A Mach 9.6	X-51A maintains Mach 5.6 for 210 secs	Hypersonic airliners
X-15 rocket plane achieves Mach 6.7				
1950s–1960s	1970s–1980s	2000–2010	2014–Present	Mid-2030s

HYPERSONIC PROMISE

▶ **GE is using ramjet technology to up the ante in hypersonic flight.** GE Aerospace

▼ **In 2025, Hermeus operated its first Quarterhorse Mk 1 flight.** Hermeus

aircraft, but at higher speeds it bypasses the incoming air around the turbine and the ramjet takes over completely.

The Chimera propulsion system will use a pre-cooled Pratt & Whitney F100 fighter engine in conjunction with an integrated ramjet to power future hypersonic flight. Pre-cooling intake air before compression allows a standard engine to operate at higher speeds with greater efficiency and reduced performance degradation. The effect can be accomplished using a heat exchanger and a cryogenic fuel, such as liquid hydrogen. These efficiency gains are needed to bridge the speed gap between the capabilities of a standard jet engine and the ramjet propulsion system needed to reach hypersonic speeds.

Hermeus explained that most hypersonic platforms are powered by a rocket engine, which limits operability, maintainability and reliability. By making an air-breathing hypersonic engine that does not require a rocket to accelerate it is setting the stage for operational hypersonic flight – meaning aircraft that can be rapidly turned around a modern aircraft. In November 2022, Hermeus demonstrated turbojet-to-ramjet transition within Chimera, which is one of the most important technological challenges enabling operational hypersonic flight.

Chimera's innovative design enables the aircraft to operate efficiently from take-off to hypersonic cruise, combining the benefits of traditional turbojet engines with the raw power of a ramjet. By using a single engine architecture for both low and extremely high speeds, Hermeus sidesteps the complexity that has long plagued hypersonic flight.

Hermeus has adopted a revolutionary approach to developing new platforms and with significant strategic partnerships the company is poised to play a pivotal role in this emerging hypersonic space. Beyond commercial aspirations, it is also working closely with agencies like the USAF to explore defense applications for hypersonic platforms.

GE ramjet technology

When considering hypersonic flight, it is worth remembering that this technology is still in the early stages of development, particularly with regard to passenger transport. Elements that will eventually go into the hypersonic mix are being designed, tested and improved to validate or rule out new technologies. Achieving hypersonic flight efficiently will depend on engine technology and that's why aerospace giant GE Aerospace (GE) is active in this field.

In July 2024, GE announced it had successfully demonstrated a new, cutting edge hypersonic dual-mode ramjet which could enable high

HYPERSONIC PROMISE

speed flight and longer range across numerous multi-mission aircraft. It also represented another key milestone for GE across its diverse portfolio of hypersonic programs.

GE had commenced testing the dual-mode ramjet in March 2024 in its clean air, continuous flow, high speed propulsion testing facility in the US, a mere 11 months after the launch of the design programme. GE reported that the testing delivered results that exceeded performance expectations and demonstrated robust operation of the dual-mode ramjet with a threefold increase in airflow compared to previously flight tested hypersonic technology demonstrators. Amy Gowder, president and CEO of Defense and Systems at GE, said: "The rapid progression from design to testing underscores our commitment to driving innovation in hypersonic technologies. This milestone not only shows the exceptional talent and dedication of our team but also reaffirms our position as a leader in the pursuit of hypersonic flight."

This solid result paves the way for the next phase of development, which will focus on continued testing and technology demonstration in alignment with GE's roadmap for integrated high speed propulsion solutions.

Sense of urgency

In January 2025, GE said it was "using ramjet technology to up the ante in hypersonic flight" and that its engineers were steadily solving the problem of hypersonic flight, allowing aircraft to safely shoot through the skies at five times the speed of sound. The GE announcement added that as engineers figure out how to bend time in passenger travel they are attending to an even more urgent priority of harnessing the extraordinary capabilities of hypersonics in the name of global security. Dean Modroukas, general manager for hypersonics, said "our adversaries have pushed the industry into an upswing" in reference to the world's current geopolitical landscape. He added that the race was on to develop hyper-efficient propulsion systems that can travel much farther and much faster, dramatically compressing an adversary's time and room to manoeuvre.

While all this sabre rattling is interesting, its relevance to hypersonic passenger aircraft is limited, apart from the fact that the high-level research will help develop the engines needed for future commercial hypersonic airliners.

Moving on from politics, GE commented that, after a year of ground testing that had successfully moved the programme towards its goal of hyper-efficient, high speed, long range hypersonic flights, the engineers were now preparing to prove their breakthrough technologies in the skies.

Hypersonic tech at GE

As touched on previously, the air-breathing ramjet engines that enable hypersonic flight are very different from the turbojets and turbofan engines on passenger aircraft. Instead of using moving components like compressors and turbines, ramjet engines depend on specially engineered air inlets that compress the air to the right pressure for combustion. A standard ramjet propulsion system ignites only when the vehicle achieves supersonic speeds that are greater than Mach 3. This is why ramjets have traditionally relied on rocket boosters or special turbo-ramjet engines to achieve the extra throttle for ignition.

GE has been working with NASA and other research partners for more than a decade to make ignition happen using pulsed detonation engine technology, which burns fuel in waves of small explosions instead of standard combustion. In 2023, GE had demonstrated a hypersonic dual-mode ramjet rig that harnessed rotating detonation combustion technology within a supersonic flow stream. This technology generated higher thrust from a smaller engine size and weight to produce significant efficiency gains. GE said it could propel aircraft seamlessly across a range of Mach numbers. The next stage of development will be key to getting the ramjet technology airborne.

▲ Chimera will use a pre-cooled P&W F100 engine with an integrated ramjet.
Hermeus

▼ Testing of Quarterhorse aircraft is done at Edwards Air Force Base, California.
Hermeus

Making a Profit

The second coming of supersonic passenger travel is offering OEMs and airlines the opportunity to be both fast and profitable.

MAKING A PROFIT

At this early stage of the revival of supersonic travel, almost all the news and insights are coming from the OEMs, but that is likely to change because it is actually their customers that will largely determine the sector's success or failure. The OEMs will have to convince airlines that there is a profitable future in establishing supersonic passenger services and that their aircraft concept is the one that will deliver results. This is why gaining the rights to fly supersonically over land is so important, particularly in the US, where transcontinental flights are highly popular and have large passenger markets.

Boom Supersonic and Spike Aerospace have laid out the expected cabin configurations for their first aircraft types. Just as airlines build new routes with smaller jets and grow from there, the supersonic OEMs acknowledge that their best chance of success is by targeting the passenger markets most willing to put convenience over price.

Assuming that new entrants can meet the updated noise regulations that ICAO is developing for supersonic flights over land, their airline customers will have the world on their doorstep and can target tourism, business and friends and family markets. It will be an exciting time to see which airlines take up the challenge, where they perceive opportunities and what routes they will select to open their new supersonic future.

There is a common misconception that Concorde was a totally loss-making venture. In reality, it made substantial profits for British Airways, both on its flagship UK-US routes and the charter market. Air France and the French government followed an alternative path with their supersonic accounting and got a different result. With a tighter grip on development costs and a greater willingness to form collaborations with partners, the second coming of supersonic operators will start off in better financial health than their predecessors.

Aircraft production capacity

As the most advanced supersonic start-up, Boom's Overture supersonic airliner is planned to carry 60-80 passengers at twice the speed of sound running on up to 100% sustainable aviation fuel with certification expected before the end of this decade.

In June 2024, Boom announced that it had completed construction of its Overture Superfactory, the first supersonic airline production facility in the United States. Located at the Piedmont Triad International International Airport in North Carolina, the campus will include a delivery centre where companies such as United Airlines, Japan Airlines and American Airlines will

▼ **United Airlines has orders and options for up to 50 Overtures.**
Boom Aerospace

MAKING A PROFIT

▲ **United Airlines is one of Overture's launch customers.**
Boom Aerospace

▶ **Passenger experience is a focus of Boom's Overture.**
Boom Aerospace

be among the first to receive their Overture airliners.

Boom said that hundreds of millions of passengers will fly supersonic on aircraft produced at the Overture Superfactory and the first assembly line has the capacity to produce 33 Overture aircraft per year, valued at more than $6 billion. Boom also plans to build an additional assembly line that will scale the facility to produce 66 Overture jets annually.

With the Superfactory construction now complete Boom will focus on operationalising the production floor. In partnership with tooling supplier Advanced Integration Technology, Boom will begin procuring and installing tooling into the Superfactory, beginning with an advanced test cell unit. As the first major piece of equipment to be installed the test cell will be used to develop manufacturing processes, optimise the flow of the assembly line and prepare staff for Overture production. Blake Scholl said: "Construction of the Overture Superfactory represents a major milestone toward ensuring the United States' continued leadership in aerospace manufacturing. Supersonic flight will transform air travel and Overture provides a

MAKING A PROFIT

subsonic jets in the up to 150 seat market and, in 2024, it delivered 73 aircraft, including 47 of its new generation E2 jets and 26 E1 aircraft. Boom's plans to increase production to 66 Overture airliners closely aligns with Embraer's annual production rate and paints a realistic picture of how supersonic aircraft demand and manufacturing will start with Boom already leading the charge.

Other types

The other prominent entrant into supersonic passenger transport is Spike Aerospace, which plans to reintroduce supersonic flight for commercial and private use. Its flagship aircraft, the Spike S-512 Diplomat, is a low-boom supersonic bizjet designed to fly quietly over both land and water, reducing international travel times by up to 50%. This is a different business model to Boom's larger capacity Overture, but one that is likely to quickly gain traction given the needs and travelling habits of international corporate, government and high net worth individuals. The ability to travel securely and privately between cities such as New York and Paris in less than four hours will attract significant demand in a sector where cost is a secondary concern compared to convenience. It is still early days for Spike, but judging by the cabin images already released the Diplomat will be a luxurious aircraft with a variety of cabin configurations.

Several other airlines and aviation companies have expressed interest in future supersonic developments, though without publicised firm orders. Carriers in the Middle East and Europe, including Virgin Group, have signalled enthusiasm for high-speed travel, often entering into partnerships, letters of intent or option agreements with manufacturers.

OEM profitability

Entering any new aviation market is a daunting experience, but when it involves something as fraught as supersonic passenger transport it is definitely not an endeavour for the fainthearted. One can only admire those who are taking on that challenge and navigating the maze

▼ Concorde was built for speed more than cabin amenity.
allthelovekatxx/ Shutterstock

much needed innovative alternative for airlines across the globe."

As of June 2025, Boom's order book stood at 130 aircraft. The very first agreement had come in 2017 when Japan Airlines announced an option for 20 aircraft. As part of the agreement, JAL made a strategic investment of $10 million in Boom and was collaborating with the company to refine the aircraft design and help define the passenger experience for supersonic travel. In June 2021, United Airlines became the first US airline to sign an aircraft purchase agreement for 15 Overture airliners with an option for 35 more. And in August 2022, American Airlines placed a deposit for up to 20 aircraft with an option for 40 more, which would position the airline as operating the world's largest supersonic fleet.

These 130 commitments represent close to four years of production on the initial assembly line at the Overture Superfactory. As the aircraft moves closer toward certification and entry into service, it is highly probable more airlines will make purchase commitments, particularly if new regulations allow supersonic flights over land.

Brazil's Embraer is currently a leading manufacturer of commercial

www.key.aero 97

MAKING A PROFIT

of government regulations, new aircraft and engine technologies, building completely new supply chains, financing the project, meeting strict environmental expectations and finding airline customers.

However, the clean-sheet revival of supersonic travel gives these innovators the opportunity to ensure they are profitable, particularly given the size of the proposed new jets and the affluent markets they are aiming to attract. It also allows for airlines or operators to differentiate their aircraft and services to emphasise what their customers want, be it more luxury, onboard workspaces, sleeping facilities or just the thrill of breaking the sound barrier.

Boom has said its initial manufacturing capacity will produce 33 Overtures at a total value of more than $6 billion, which amounts to around $180m per aircraft (in comparison, the list price for a Boeing 737 MAX 8 with capacity for up to 200 seats is around $120m).

Given the meticulous planning that has gone into developing the Overture there is no reason to think Boom is not making a healthy profit on each aircraft, although how it accounts for the huge development costs may affect profitability in the early years.

▼ Japan Airlines is an Overture customer and a Boom strategic partner.
Japan Airlines

Airline perspectives

Reviving supersonic flight in the smaller scale envisioned by Boom will present existing airlines with plenty of opportunities to test the water and open the door for niche operators to establish small fleets and specialise in supersonic services. There is no reason for airlines not to be profitable given the elasticity of pricing in this high-end market and flying from New York to Paris in less than four hours on relatively small aircraft will attract more than enough demand. There has not been any indication of ticket pricing, but in the initial period it will be priced at a premium, although anything beyond that is pure speculation at this early stage. Airlines will also need to make significant investments in areas such as training for flight and cabin crew, marketing, maintenance and ground handling, but as Concorde proved, all of that can be managed without too much fuss.

Another factor to consider are the long waiting lists for new-generation subsonic narrowbody aircraft and the delays that seem to be endemic in that sector at both Airbus and Boeing. Some of the Middle East premium carriers already have private jet operations and would look at supersonic travel from Dubai, Doha or Abu Dhabi as an open

MAKING A PROFIT

▲ American Airlines will have the world's largest supersonic fleet.
American Airlines

market for highly profitable business, government and head of state opportunities. With Boom confident it will have Overture certified by the end of the decade and entry into service soon after, airlines who want to enter the supersonic sector will have to show their hands, if for no other reason than to secure production slots.

The Spike model of business jet travel could be easier for operators to establish, where the concept of high-end cabin configurations and privacy are already the benchmark. What is being added is what these customers crave and that is getting where they want to be when they want to be there, but faster.

The takeaway from all this is that supersonic passenger services will not come back as Concorde imitations but as a new way of harnessing the power and attraction of speed to connect cities that currently require a long-haul flight. The turning point for profitability will be if and when supersonic flights over land get the green light, opening up lucrative markets across North America, Europe, Asia, Australia, Latin America and other regions worldwide.

MAKING A PROFIT

▶ The Spike Diplomat is offering a luxurious and faster flight.
Spike Aerospace

▶ Spike's S-512 Diplomat bizjet offers numerous cabin configurations.
Spike Aerospace

▼ Boom's production line has a similar capacity to what Embraer delivers annually.
Embraer

Attracting customers

It is fair to say that while Concorde offered speed, the cabin was not a premium experience, especially by today's standards for first or business class passengers. Spike has put out a variety of cabin concepts that closely resemble high-end business jets, with the added touch of interactive screens instead of windows.

With its larger cabin and passenger numbers, it will be interesting to see how Boom configures its 60-80 seat interior and if its customers will create separate cabin classes within the airliner. This is a clean sheet for airlines to put their stamp on supersonic flight and how they complement it with exclusive airport lounges and other premium passenger touches, much as British Airways and Air France did so successfully all those years ago.

The attraction of flying supersonic is not only about the speed, but also the unique experience of prestige, comfort, time saving and making connections in quick time, such as a weekend in Paris for US travellers. The exclusive feel of breaking through the sound barrier adds to an experience that could be strong enough to create a loyal, high value and expanding customer base.

Just as Concorde exuded glamour and luxury, tomorrow's supersonic airlines have the opportunity to market themselves as unique aspirational travel experiences where passengers feel part of a special band of adventurers. The limited number

MAKING A PROFIT

▲ Boom has orders and commitments for 130 Overtures.
Boom Aerospace

◄ Embraer delivers a similar number of aircraft to Boom's planned capacity.
Embraer

of aircraft and seats will only add to this exclusivity, thus achieving profitability seems more certain than what Concorde faced when it started service close to 50 years ago. Of course, there is still considerable ground to cover until launch and, as we saw with the abrupt demise of Aerion, nothing is guaranteed until the aircraft are certified and operating safely and reliably meeting schedules.

Only time will tell how financially successful this second iteration of supersonic travel will be for OEMs and their customer airlines and operators, but the measured pace of development and the changing political will to enable it are in contrast to the first time around.

◄ Qatar Airways is likely to show interest in the Spike Diplomat.
Qatar Airways

SUSTAINABLE SUPERSONIC

Sustainable Supersonic

The next generation of supersonic passenger airliners will be greener and more sustainable than Concorde, but will that be enough for the environmental lobby?

Concorde and sustainability are not often mentioned in the same sentence, but neither would be any other form of transport conceived and designed in the 1960s. Automobiles, buses, trains, commercial ships and trucks were all gas guzzlers and prolific emitters until environmental responsibility started to rein them in a decade or so later.

Neither supersonic or subsonic aviation was any different, but the sector has made concerted efforts in the last decade to reduce fuel use and carbon emissions. At the 77th IATA Annual General Meeting in Boston, Massachusetts, on October 4, 2021, a resolution was passed by IATA member airlines committing them to achieve net-zero carbon emissions from their operations by 2050. This pledge brings air transport in line with the Paris Agreement's global temperature goals. Reaching this ambitious challenge will require the co-ordinated efforts of the entire industry, including airlines, airports, air navigation service providers, manufacturers and government bodies.

▼ Concorde was designed decades before sustainability was an issue. ODIN Daniel/Shutterstock

SUSTAINABLE SUPERSONIC

While figures differ on just what percentage of global CO_2 emissions are produced by aviation, the IATA figures reported that in 2024 the sector was responsible for 2.05%, with international services accounting for 1.3% and domestic travel for 0.7%. In comparison, road transport produced 11%, power generation 26%, agriculture 11% and fuel production 10%.

While laws are being passed in some countries to ban flights on short sectors, the reality is that aviation has done more than almost any other major industry to reduce its emissions, thanks largely to engine OEMs like Pratt & Whitney, GE Aerospace, Rolls-Royce and CFM International, whose new-generation powerplants have cut fuel burn and emissions by around 16-20%.

Supersonic challenges

Many have lamented the more than 20-year gap between the retirement of Concorde and the nascent revival in supersonic passenger travel, which will not see a new aircraft certified and in service before the end of this decade or early in the next.

However others believe that gap will enhance the success of the next generation of supersonic airlines as it will showcase what new technologies can offer. New entrants will have much more realistic expectations of the type of aircraft needed to be both profitable and environmentally sustainable.

The reality is that the physics of flight at speeds above Mach 1 sharply increases wave drag, which demands more thrust from the engine and therefore more fuel to maintain cruise speed. Fuel consumption on Concorde was approximately three to five times that of a Boeing 747 on a similar route. While modern engine designs have improved that figure, there is still a significant gap between subsonic and supersonic airliners.

Burning more fuel directly translates into increased CO_2 emissions which, like sonic booms or airport noise, will not be welcomed in today's cleaner environment as supersonic air travel fights to push its green credentials.

Cleaner footprint

Current plans for new aircraft such as Boom Supersonic's Overture or the Spike Aerospace S-512 don't include any widebody jets, focusing either on business or regional aircraft. Also, in the early stages, the number of supersonic aircraft flying will be limited and unlikely to significantly impact aviation's total CO_2 emissions. That's not to say that aircraft designers and engineers are deaf to these issues. By sourcing and validating stronger and lighter materials that reduce weight and employing advanced aerodynamic technologies they are making huge advances to reduce fuel usage and CO_2 emissions to much lower levels than those posted by Concorde or the Tupolev Tu-144.

While supersonic flying is unlikely to be ever classified as 'green' there are ways it can become more environmentally responsible than previous generation supersonic aircraft, through improved fuel efficiency, cleaner combustion and lower non-CO_2 effects, and by significantly reducing operational and sonic boom noise.

Fuel efficiency

One of the main issues with supersonic engines is that they have typically been pure turbojets or low bypass designs that are optimised for high-speed thrust but are inefficient at subsonic speeds. As discussed in the chapter on engines, that is changing with manufacturers producing bespoke powerplants rather than using derivatives of engines already available from the major OEMs.

Variable-cycle turbofans, such as the GE Affinity that was destined for the now cancelled Aerion AS2 supersonic aircraft, are designed to shift between a low-bypass, high-thrust mode and a higher bypass fuel efficient mode for take-off and subsonic segments of the flight, which improves the overall efficiency and fuel burn of the aircraft.

On the materials side, ceramic matrix composites and advanced titanium alloys allow engines to operate at higher temperatures with less cooling air needed to improve thermodynamic efficiency. New-generation engines, such as Pratt & Whitney's geared turbofan (GTF) and CFM International's Leap series are doing exactly that on subsonic narrowbodies like the Airbus A320neo and the Boeing 737 MAX family of passenger aircraft.

Another technology that is being tested and validated for supersonic aircraft engines is the use of variable

SUSTAINABLE SUPERSONIC

▲ New-generation subsonic engines have cut fuel use by 16-20%.
Pratt & Whitney

▶ CFM International's LEAP engine is an emissions reducer.
CFM International

geometry inlets that can precisely manage airflow at supersonic speeds. This reduces shockwave losses and wave drag, which means that less thrust and fuel is required to maintain cruise, leading to less harmful emissions.

Cleaner combustion

As supersonic aircraft will be flying at altitudes around 50,000-60,000ft, the non-climate effects of nitrogen oxide (NOx) emissions and water vapour also present challenges to reducing the per-passenger environmental impact. Designers are working to significantly reduce NOx and soot, to directly mitigate some of the most damaging atmospheric effects of high-altitude supersonic flight.

As the name suggests, NOx emissions contain a group of highly

SUSTAINABLE SUPERSONIC

reactive gases containing nitrogen and oxygen primarily nitrogen oxide (NO_2) and nitric oxide (NO). These gases are significant air pollutants and have detrimental effects on human health and the environment, with NOx emissions primarily originating from the combustion of fossil fuels in engines and industrial processes.

NOx can catalyse the destruction of ozone (O_3) in the stratosphere, which protects Earth from harmful ultraviolet radiation. Studies from the Concorde era suggested that widespread supersonic travel could cause measurable ozone layer depletion. This is something of which designers need to be particularly mindful, as it will be seized upon by climate activists when supersonic passenger transport returns in the 2030s.

Non-CO2 effects

Flying at 50,000-60,000ft amplifies the climate effects of NOx and water vapour, so engine makers are introducing their own set of innovations to help reduce this. Modern subsonic and supersonic engines are already designed to operate on 100% sustainable aviation fuel (SAF), which can be refined to burn cleaner, producing less soot and fewer contrail-forming particles.

Lean burn combustors reduce peak flame temperatures to lower NOx emissions that are especially

▲ Major airlines like Emirates have proven SAF to be effective alternative. Neste

▼ Airbus A321neo aircraft have cut airlines fuel bills by around 16%. Airbus

SUSTAINABLE SUPERSONIC

▶ Neste has recently expanded SAF production in Singapore. Neste

harmful to ozone at supersonic altitudes. New-generation engines are more efficient at mixing fuel and air through the use of improved injector designs that ensure a more complete combustion process that reduces unburned hydrocarbons and unburned particulate emissions.

Subsonic sustainability

Nearly four years have passed since most of the world's airlines committed to IATA's pledge to make commercial aviation a net zero emissions industry by 2050. When making those commitments, some of the larger airlines set ambitious interim goals for 2030, but we are seeing many airlines dial back on those goals even though they have five years to make up for lost ground.

While there are a number of opportunities for airlines to decarbonise, the real gains can only come from their operational activities, much of which relies on outside organisations. Over the last few years, it has become rather boring to read announcement after announcement advising that another airline has successfully operated flights using a blend of SAF and regular jet fuel. There is no question that today's aircraft can run on SAF blends of up to 50%, and technically most can already run on 100% SAF if it were available at a competitive price. It is not about the airlines, the engines or the aircraft – the dominant issue is that there is simply not sufficient SAF available to start making a significant difference to net zero goals.

In 2024, global jet fuel demand reached around 330 million tonnes, and while the production of SAF doubled from 2023 levels, IATA said it accounted for just 0.3% of global jet fuel production. Airlines around the world are scrambling to lock in SAF supplies, but with minimal new investment in SAF production there isn't enough to go around. In December 2024, IATA's director general Willie Walsh observed: "SAF volumes are increasing, but disappointingly slowly. Governments are sending mixed signals to oil companies, which continue to receive subsidies for their exploration and production of fossil oil and gas. And investors in new-generation fuel producers seem to be waiting for guarantees of easy money before going full throttle. But make no mistake – airlines are eager to buy SAF and there is money to be made by investors and companies who appreciate the long-term future of decarbonisation."

IATA analysis showed that to reach net zero CO_2 emissions by 2050, between 3,000-6,500 new renewable fuel plants will be needed and the annual average capex needed to build the new facilities over the 30-year period is about $128 billion per year. This is significantly less that the $280 billion per year being invested in solar and wind energy.

The world's largest producer of SAF is Finnish company Neste, which has made significant investment in its Singapore and Rotterdam facilities to increase output. Despite this, Neste will only have the capacity

▼ Rolls-Royce has conducted its own validation of SAF. Rolls-Royce

SUSTAINABLE SUPERSONIC

to produce 1.5 million tonnes of SAF in 2025, compared to the 2024 global usage of jet fuel sitting at 330 million tonnes. Neste MY SAF is already available in San Francisco, Los Angeles, Frankfurt, Amsterdam, Singapore and Tokyo Narita. These and other airports globally have already proven they can run SAF through their normal refuelling systems, such as at Singapore Changi Airport where Singapore International and Scoot operate scheduled flights refuelled with MY SAF.

As Willie Walsh highlighted, the issue with SAF is not technical but logistical. It has taken nearly 100 years to develop the global refining capabilities and supply chain for the automobile industry, yet aviation seems to expect that energy suppliers will do the same for aviation in 30 years or less. To spend the billions quoted by IATA to build new production facilities, investors need to be sure they will get their money back and, without government support or mandated use of SAF, those guarantees are not there. Investors are also spooked by the talk of alternative technologies such as hydrogen or electric power being the preferred long term solution to aviation decarbonisation.

The other aspect that is finally dawning on the aviation industry is that the supply chain for procuring the feedstock for SAF production has not been thoroughly investigated. There are an endless range of biological sources from which SAF can be produced, but converting agriculture away from food production to producing feedstock for SAF does not appear to be a favoured option.

For the last five years, the aviation industry has pinned its hopes on SAF, particularly for long-haul airlines where using electric or hybrid-electric power is not an option. With production expanding at dangerously slow levels, it's no wonder that airlines such as Air New Zealand have announced they will not meet their interim sustainability goals due to this uncertainty surrounding SAF supply.

Commercial supersonic operators will likely be able to procure SAF, because their premium fares will be able to cope with the higher priced fuel and because fleet sizes will be small in the early years of supersonic passenger transport. However, this supply issue is a real concern to all of commercial aviation, as the industry needs SAF to bridge the gap until more innovative solutions are found.

Future outlook

In 2025 there are no modern supersonic airliners to scrutinise, so predictions for the future are highly speculative. As with subsonic aviation, the promise of SAF looms large, but without it, it will be very difficult for the first batch of new supersonic aircraft to meet society's environmental expectations. Conversely, if Concorde is used as a benchmark, these new aircraft will be miles ahead in terms of efficiency, fuel burn, carbon and NOx emissions than the original supersonic marvel. If SAF becomes available at scale, the figures will improve further.

▼ SAF has proven itself in trials globally. Neste

SUSTAINABLE SUPERSONIC

▲ **Malaysia Airlines is another airline to have fuelled flights using SAF.**
Malaysia Airlines

▶ **Most of the environmental gains have come from new-generation engines.**
ST Engineering

SUSTAINABLE SUPERSONIC

◀ Supersonic airliners need to use SAF to get their green credentials.
Malaysia Airlines

ICAO and aviation regulators will ensure that all supersonic jets meet the upcoming noise standards around airports and that the sonic boom is reduced to such a level that causes no disruptions to communities under and near to the over land flightpath. While Concorde faced legal action and public protests, new entrants will know in advance what's required of them and adapt their designs to meet those rules.

Advances in engine efficiency, sustainable fuels and noise reduction are a promising start, but significant sustainability challenges persist that must be addressed if supersonic passenger transport is to reach its full potential. It will also need a co-operative approach from engine and airframe manufacturers, airlines, regulators and politicians to reach its potential and ensure that the pursuit of speed and convenience does not hinder aviation's 2050 net zero commitments.

Supersonic aircraft are unlikely to ever be as energy efficient as the best subsonic design. Even with all the advances talked about above, it is unlikely that supersonic airliners will be less energy efficient than the latest generation of Airbus A350 and Boeing 787 widebodies. The opportunity open to supersonic operators is not to match subsonic efficiency, but to be carbon neutral by using SAF and minimising non-CO_2 effects.

▼ SAF flights have been plentiful in recent years.
Malaysia Airlines

▲ New supersonic concepts like the Spike S-512 are on their way.
Spike Aerospace

Supersonic in 2050

As the world becomes more connected, the desire for speed will only intensify.

SUPERSONIC IN 2050

▶ Overture's BoomLess Cruise will refract sonic booms upwards. Boom Aerospace

▶ Projections on the future of supersonic passenger transport.

▼ The X-59 launch at Lockheed Martin Skunk Works in 2024. NASA/Steve Freeman

In June 2025, Airbus released its latest Global Market Forecast, which showed that the international commercial airline fleet will almost double in the next 20 years, with a higher proportion of new aircraft replacing older, less fuel-efficient airliners. At the end of 2024, there were 24,730 aircraft in the global fleet and that will increase to 49,210 by the end of 2044, with a demand for 43,420 new aircraft. Of these, 34,250 (79%) will be narrowbodies and 9,170 widebodies (21%). Airbus also forecasts that world air traffic will grow at a mid- to long-term compound annual growth rate of

THE FUTURE OF SUPERSONIC PASSENGER TRANSPORT

Low-boom business jet prototype	First generation SSTs enter service	Overland routes open	Integration into premium market
Technology Breakthroughs Low-boom design Sustainable aviation fuels	**Market Phases** Premium routes Small fleets High fares	**Market Phases** Overland routes Growing fleet	**Market Phases** Hundreds of aircraft in service
2025	2030	2035	2050

3.6%, with stronger traffic growth in Asia and the Middle East, led by India and the Peoples Republic of China. This means nearly 20,000 new aircraft will be delivered to the Asia-Pacific region and China.

The relevance of this to supersonic passenger services is that Airbus believes nearly 80% of new aircraft over the next 20 years will be short- to medium-haul models serving routes that would be of no interest or value to a potential operator of supersonic jets. This is unsurprising, but points to the fact that supersonic services will remain niche operations for many years to come and that it will be a long time before they migrate into mass transportation, where the price of the ticket usually outweighs the time it takes to get to the destination.

SUPERSONIC IN 2050

Over time that may change, as supersonic travel builds its reputation and more aircraft are seen at airports and talked about in the media. The potential size of the market for supersonic aircraft is an estimate at best, but given the high development costs for new aircraft, any potential new entrants will need to be well financed or follow Aerion into oblivion.

Supersonic promise

Supersonic passenger travel was once predominantly the preserve of the rich and famous, but the world has moved on considerably since the demise of Concorde in 2003. There are significantly more city pairs that have direct connections and new-generation aircraft such as the Airbus A321XLR are flying long-haul routes with narrowbody economics and low seat mile costs.

After more than 20 years without civilian aircraft breaking the sound barrier, a new generation of technologies, manufacturers and regulatory possibilities are emerging. Boom Supersonic is on track to achieve certification for its Overture supersonic airliner around the end of this decade, with entry into service due early in the 2030s. If the sonic boom can either be eliminated or reduced to the sound of a gentle thump then over land supersonic routes will open up, creating more possibilities for airlines.

No manufacturer has yet built a supersonic narrowbody size passenger airliner and proven that their computer simulations will work in the real world. Boom talks about having the right atmospheric conditions as a prerequisite to BoomLess Cruise and how its small passenger-less pilot demonstrator produced no boom, but the true effectiveness of the Overture airliner will not be proven until the end of this decade.

This is why the work of NASA and its X-59 experimental aircraft is so vital to reviving supersonic flight and feeding scientific data to ICAO as it looks to develop new noise regulations for flights over land. Unless Boom, Spike and the proposed Chinese or Russian projects can meet those noise regulations and the ban is lifted, then supersonic flight will be no better off than it was in 2003.

Changing technologies

The sonic boom issue and the concepts around shaping it to cut noise on the ground have been covered in depth throughout this bookazine. It is true to say that progress is being made and early results are promising but, despite some claims, the issue has not yet been solved.

On the plus side, advances in material science are presenting designers and engineers with opportunities for innovation, incorporating materials that can better endure the stresses and strains that supersonic flight puts on aircraft. It must be remembered that airlines need reliable and safe aircraft that fly every day, so the maintenance practices around supersonic aircraft will need to develop to keep them in the air.

Advanced composites such as carbon fibre reinforced polymers and high-temperature ceramics

▲ NASA's X-59 has a unique 30ft nose.
NASA

▼ How Spike's S-512 will slash travel times from Dubai.
Spike Aerospace

SUPERSONIC IN 2050

▲ It all started with the iconic Concorde.
Jeang Herng/Shutterstock

enable stronger, lighter and more fuel efficient airframes, but also make the aircraft more resilient, reducing maintenance costs and extending service lifespans. These composite materials are produced with high temperature resins that also offer improved resistance to the aerodynamic heating generated at Mach 1.2-2.2, while titanium alloys that were used sparingly in Concorde are now more cost effectively machined for use in the high heat sections of the airframe.

Environmental concerns

New generation subsonic engines such as the Pratt & Whitney GTF and the CFM Leap are already using advanced ceramic materials to reduce fuel consumption and emissions by around 20% compared to previous generation powerplants.

The sustainability issue has quickened the development of non-afterburning turbofans with variable inlets and exhaust nozzles optimised for high-speed cruise. The goal is to maintain supersonic efficiency without sacrificing subsonic performance, allowing for greater route flexibility.

Despite producing just 2.5% of global carbon emissions, commercial aviation is facing significant pressure from governments, activists and the public to reduce emissions further. Current jet engines can already run on blends of SAF and almost all of the mainstream narrowbody and widebody powerplants are ready to shift to 100% pure SAF.

The same will be true of supersonic engines, particularly those being developed for new aircraft such as Boom's Symphony, although the drawback with SAF is not performance but availability at scale and high cost compared to current jet fuels. Electric propulsion is hamstrung by battery technology and not likely to be adopted for supersonic engines for the foreseeable future.

It is a little surprising that development programmes are not embracing hydrogen technologies, which are in their infancy now but will develop in coming years and present a clean propulsion alternative. When produced using sustainable sources, green hydrogen is the ideal answer to aviation's environmental critics, although hydrogen's low volumetric energy density, storage requirements and associated infrastructure challenges make it a more distant prospect for supersonic airliners than for subsonic jets.

Public trust

Technology is a big part of reviving supersonic travel, but equally important is gaining the public's trust and its licence to bring back a mode of travel that was seen as elitist and beyond the reach of most travellers. To be successful, operators need to develop a new model to reduce costs, through fuel efficiency, scalable manufacturing and optimising maintenance.

Supersonic passenger flights will have a profound effect on society and could change travel patterns in a similar way to the arrival of the Boeing 747 in the 1970s, which opened up a previously out of reach adventure to the masses.

The first generation of new SSTs will almost certainly serve the premium travel segment and operate on transoceanic city pairs where time savings are greatest, such as New York-London, Tokyo-Los Angeles and Sydney-Singapore. These markets have a concentration of high-yield passengers in business, government, and high-net-worth demographics where ticket prices are less of a concern.

Pathways

Over the next decade, it is highly probable that supersonic passenger transport will return, but ultimately it will be airline customers who will decide if it will be a commercial success or a failure. A new supersonic age will reshape global aviation, just as the jet age did in the the 1960s, especially if China flexes its muscle in this field.

Connecting communities in a matter of hours will redefine tourism opportunities, business schedules and political affiliations by making it possible to visit faraway destinations in a single day. While glamour largely defined Concorde, the future of supersonic passenger transport lies in convenience and turning the technology into a mainstream travel option for the many and not the few.

The dream of flying faster than the speed of sound will always capture the imagination, but its realisation will require collaboration across governments, regulatory agencies, the aerospace industry and the general public. Supersonic travel represents a leap in human possibility and as the world becomes more connected the need for speed and the desire for sustainable, accessible flight will only intensify.